N MAXWELL LANDER

CAR[redacted]ALY

ThreeL Media BERKELEY

PUBLISHED BY
THREEL MEDIA | STONE BRIDGE PRESS
P. O. BOX 8208, BERKELEY, CA 94707
WWW.THREELMEDIA.COM

PHOTOGRAPHY BY N MAXWELL LANDER
FOREWORD BY ALEX LEITCH
COVER DESIGN BY ZACHARY J BAUTISTA

LIBRARY OF CONGRESS CATALOGING-IN-PUBLICATION DATA ON FILE.

ISBN 978-0-9905571-4-2

PRINTED IN KOREA

WELCOME BACK TO T

HE SPRAWL

Max Lander's imagery lives strong in the sprawl, a mess of grit and sex and good lighting. He is an excellent photographer, concerned with the camera accurately capturing what he sees in the marks and curves of bodies living past where "nature" has a meaning. He is familiar with the uses and abuses of fluorescents, and he loves the dystopia that we were all promised in the early nineties, made in pieces: your gun, your girl, your payoff, all the themes of noir dressed up in pleather and PVC, harsh-lit to make the bodies softer. Max shoots some of the best portraits in Canada because he is interested in the details that make his subjects beautiful, what makes them stand out against their background of grime and harsh surfaces. Lander has made a career of looking tightly at the angles of faces and bodies through a lens.

The act of looking through a lens is itself a form of fetish, the spiritual elevation of tools imbued with implicit consequence. Fetish, like prestige, is a word with deep meaning. In its earliest forms, the word means "made by art" or "artificial, " given sorcerous powers by the Middle Ages, brought to English in the first wave of globalization, the colonial period, and finishing with the Victorian definition we still use: a sexual interest excited by the use of a specific tool or body part. The attention to detail is what makes it hot. Porn is about the camera paying close attention, a pure fantasy, an artifice more perfect than life can be. Good porn is when someone knows what to leave out to make the fantasy stronger. Good porn is a type of sorcery. A pipe-light left off to the side to show the curve of the hip, the bloody face of a man without feelings, who pulls bruised, pretty things out of water by their hair. The shrapnel from unseen explosions are an essential part of the aesthetic, all that skin run ragged, all those bodies twisted over a wash of kipple. No-one likes sex to be clean. Pay attention to the marks on the skin, look at what the lighting works to show you.

Now that we can compose both tools and body parts

almost at-will, fetish is popular, as are fantasies of control, but good images are still hard to make. Cameras don't see like people, with the eyes coming forward until they're all you can remember, the blood ultra-vivid even in a dark room. To make a good image, you make sure that the camera knows what part to look at, which leads to the old joke: porn is in the lighting. The lighting in these images is incredible, often hand-built. The light reveals what the lens should look at, forces the camera to pay attention to the subject at hand. Fetish photography is a specific and difficult discipline, because of the temptation to put all the weight on the tools and none on the people. People are where sex is, though, and sex is why we look at the dirty pictures.

Sex was dangerous when we were growing up. Sex with bodies always a risk, never an easy choice, always slightly fatalistic. We'd talk about having safe sex only to access the other thing, negotiate a consent held impossible to give, use tools to get to the pictures we wanted for real. Before the internet turned up, dirty pictures had to be bought, paid-for, traded around. Famously, filth would not be returned from major photo developers. Teenagers swapped stories of development houses refusing to hand over images, paid too much for Polaroid as a consequence, passed around the same magazines again and again - or maybe books, about vampires, with hand illustrations heavy on the ink. The internet gave access to the kind of images that film development houses used to refuse to hand over, even if they kept the images for private showings. A bathroom can be a darkroom, if necessary, leading to black and white being less an aesthetic than a necessity. The aesthetic comes from the tools that make the work.

In this, porn is not unlike science fiction, itself a fetish-based fiction. Science fiction is defined by its fetishes: space ships, lasers, neon, leather. In science fiction, the use of weapons and surface detail are what matter. Cyberpunk science fiction is more fetishistic than most, concerned with the state of the art. Cyberpunk has a focus on lavish-and-squalor dystopia, where we were famously not promised jetpacks: we were promised sex and dirt, and we liked it. Grimy dystopia feels nostalgic now that the cold glass Singapore future has beaten off our old Chiba City fantasies. It's easy enough to wipe a slick of blood off an encased smartphone to stare at the pictures, which are plentiful now that cameras are in everything and pictures are largely terrible. The beautiful creatures contained in this book are a past, a slice of maybe-this maybe-that promised in paperbacks on the bus, developed in a dialup connection to the internet, dreamed into a gritty world wrapped in wet-look PVC, made wetter by contrast to a backdrop of grime.

WE NO LONGER LIVE IN A WORLD WHERE WE ARE CONSTRAINED BY THE BODIES WE WERE BORN INTO, THE PORN WE GOT HANDED BY FRIENDS: WE CAN CHOOSE WHAT TO BE AND WHAT TO LOOK AT. WE CAN CHOOSE WHAT KIND OF FUTURE WE WANT TO LIVE IN AND WHAT WE WANT TO GET OFF ON. THIS BOOK IS, STRAIGHT UP, BLOOD-HOT NEON SEX. YOU ARE MEANT TO NOTICE THE BRUISES.

ENJOY.

INTRODUCTION

This book is hard to explain. It has been many things before it came to rest at what it is: a documentation of transhuman bodies, an exploration of body modification, a collection of fictional movie characters, a look at future sex, a whole lot of masked men. It is now none (and maybe also all) of them. I've used silly strung together phrases to attempt to describe it. It is "pseudo" a lot of things. Porn. Portraiture. Science Fiction. Cyberpunk. Noir. Fetish. Fantasy. Violence. It is all of these things. It is that throaty whisper. It is orange mesh. It is crash and burn. It is DIY. It is occasional romance. It is the 90's. It is the moment when you are yelling "make out!" at the TV (and they do). It is what happens when you say "Hey, I've gotten a book deal and it's gonna get weird and be hard to explain and would you like to donate your body to the cause?" (and they do).

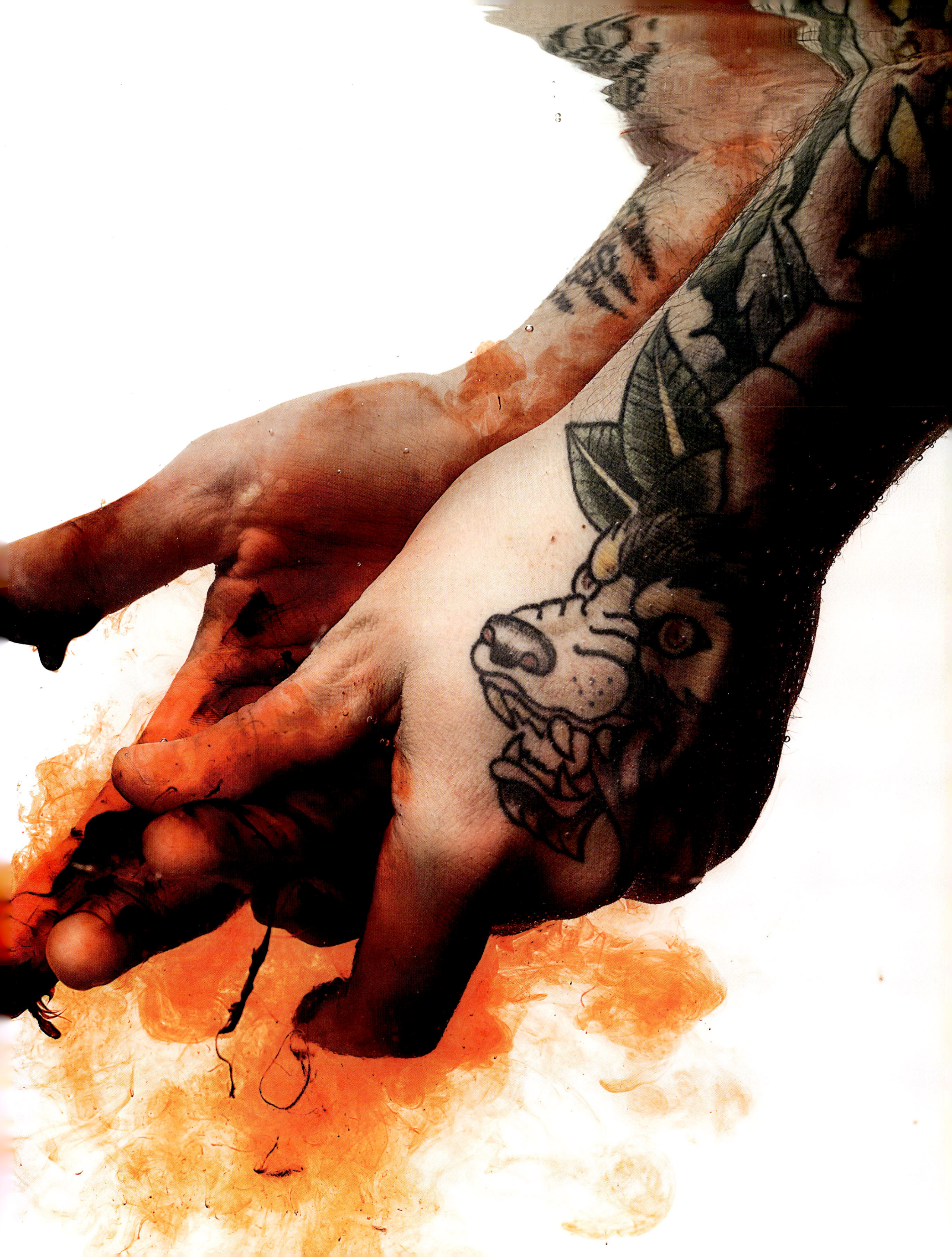

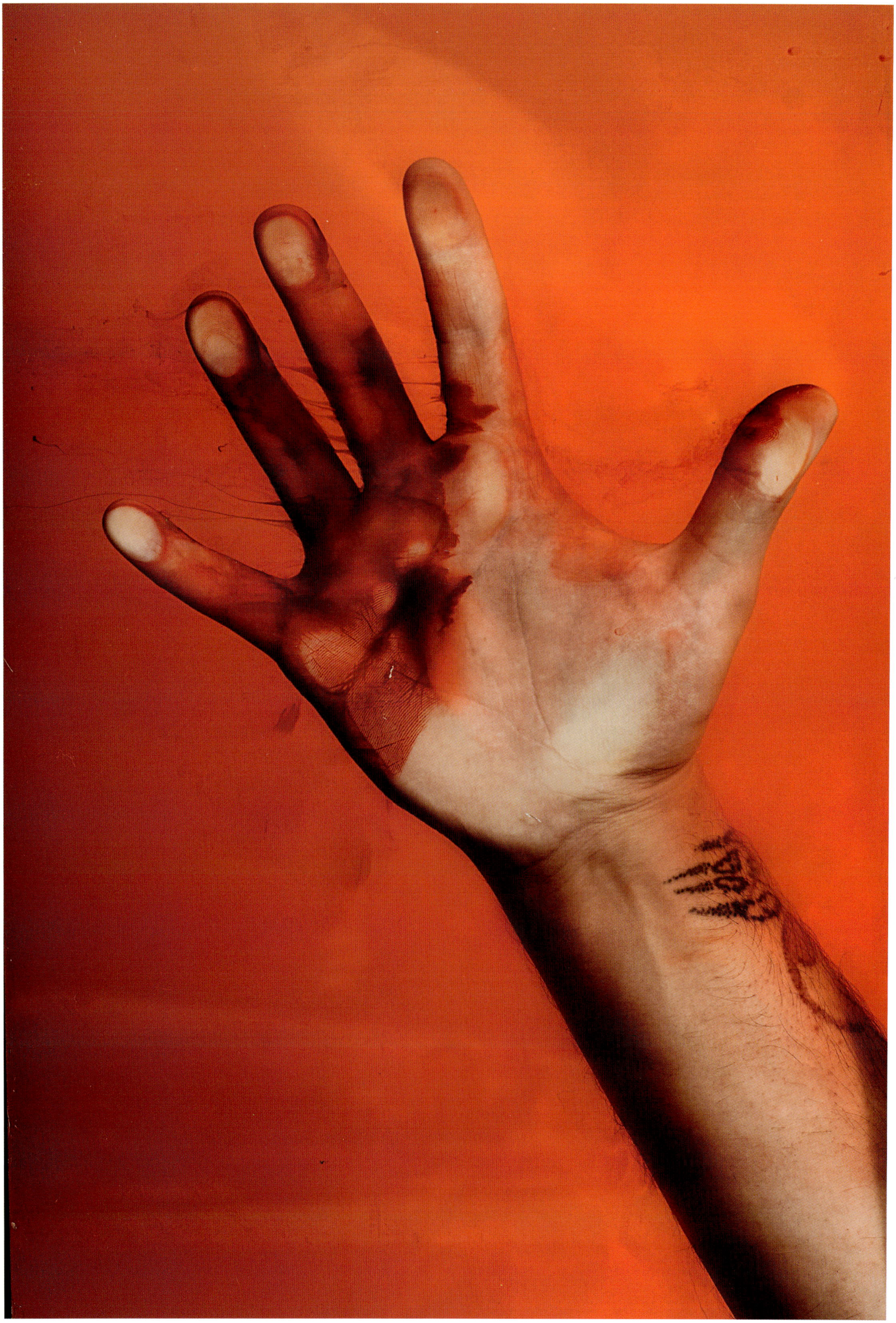

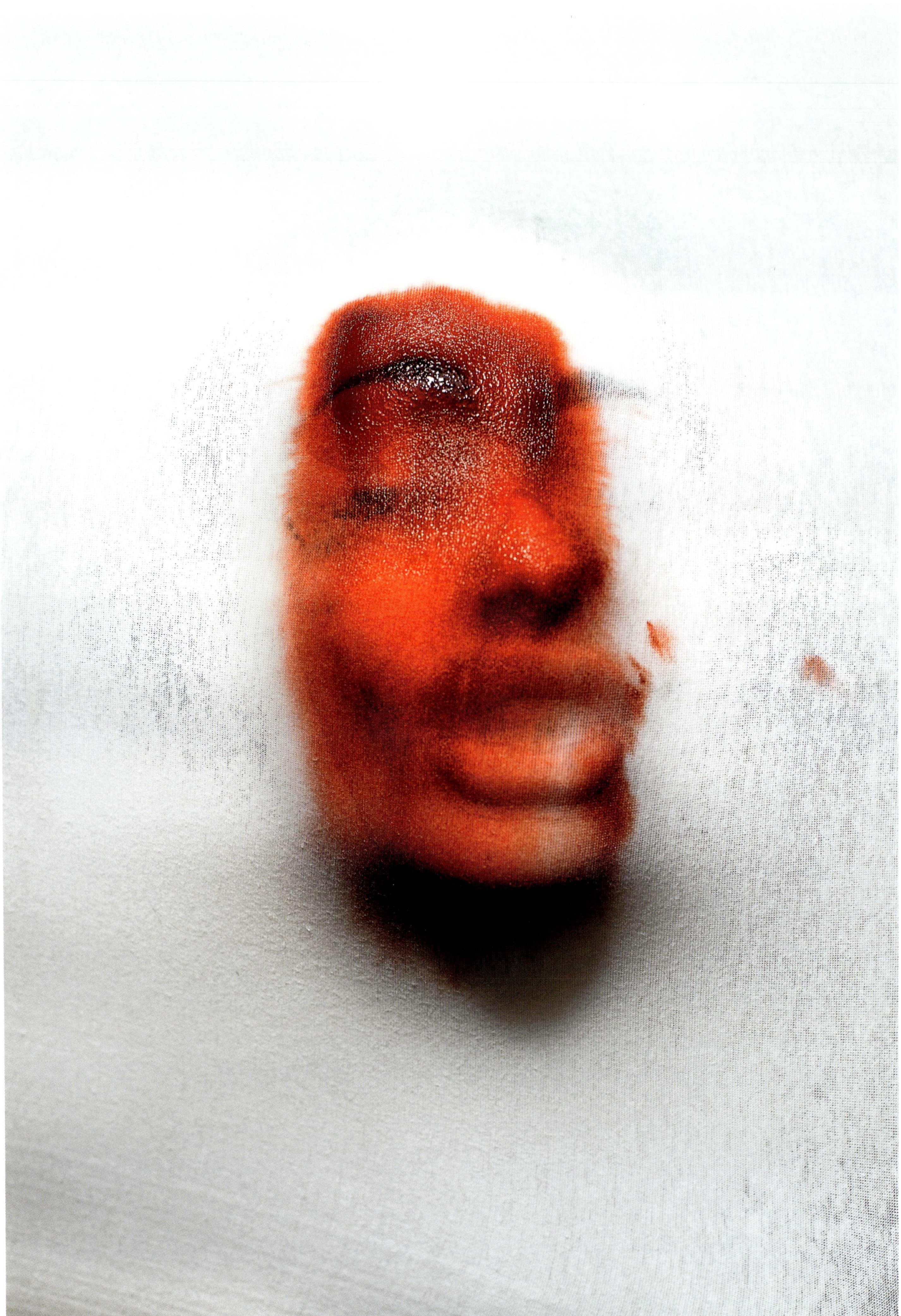

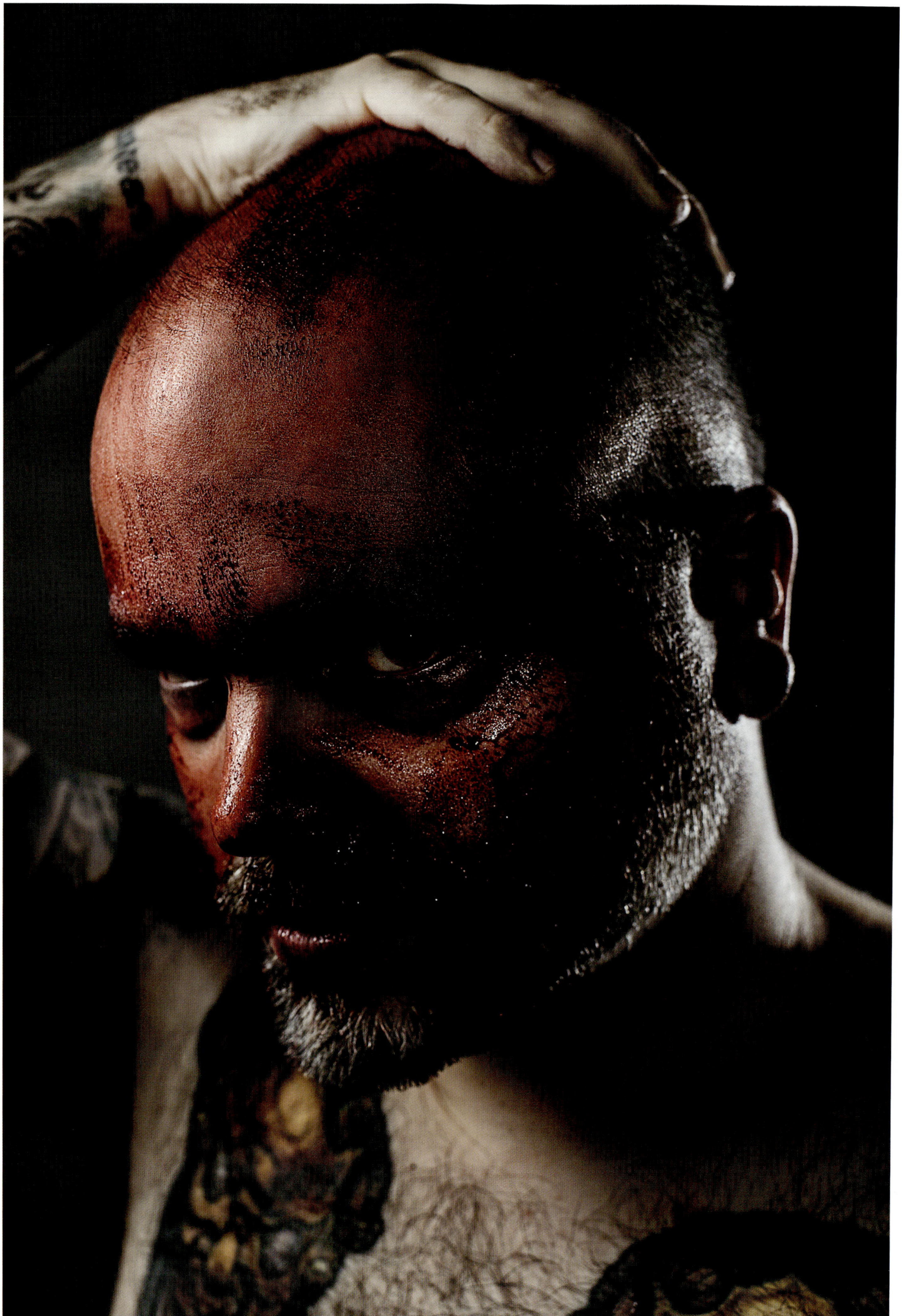

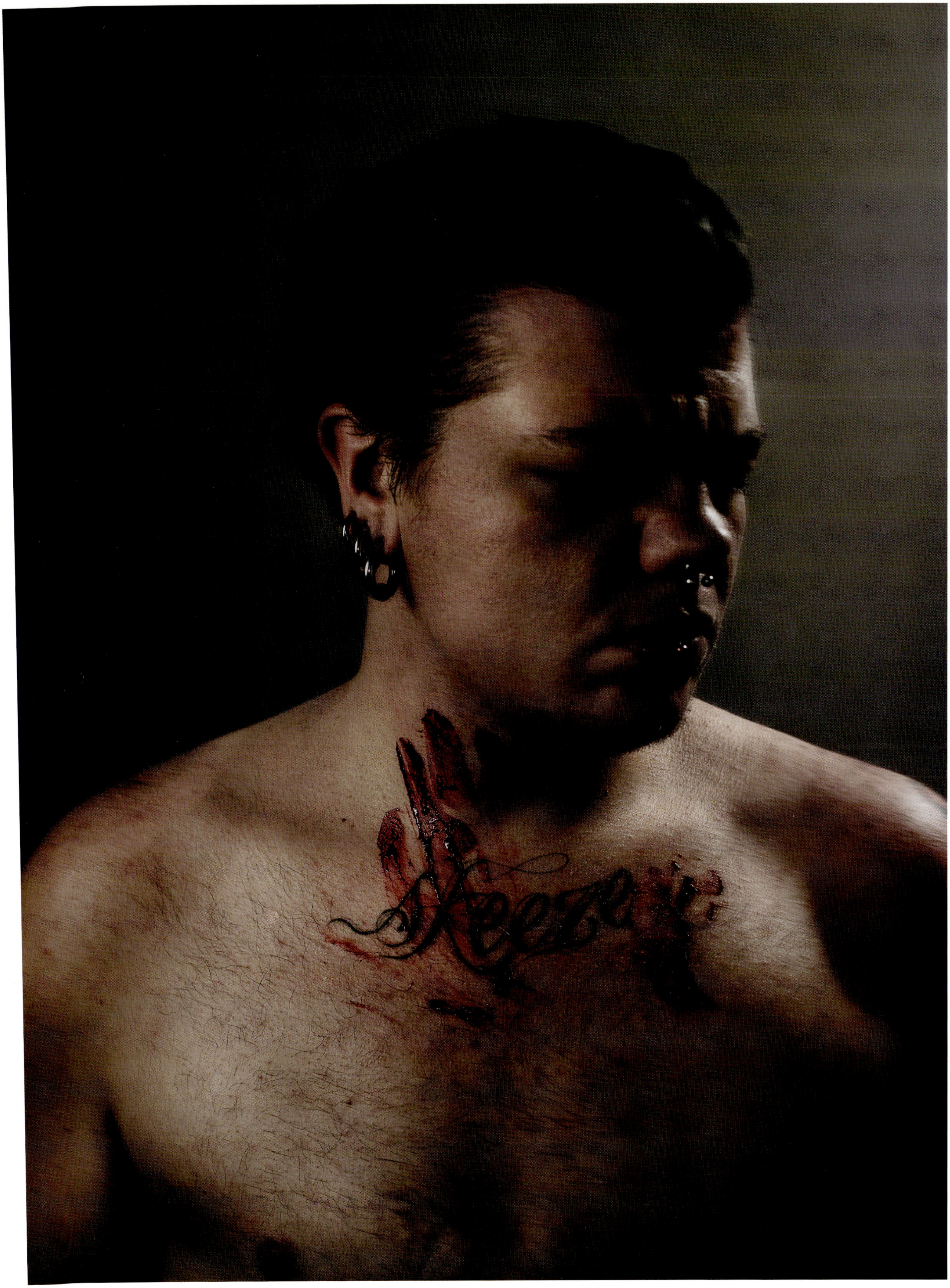

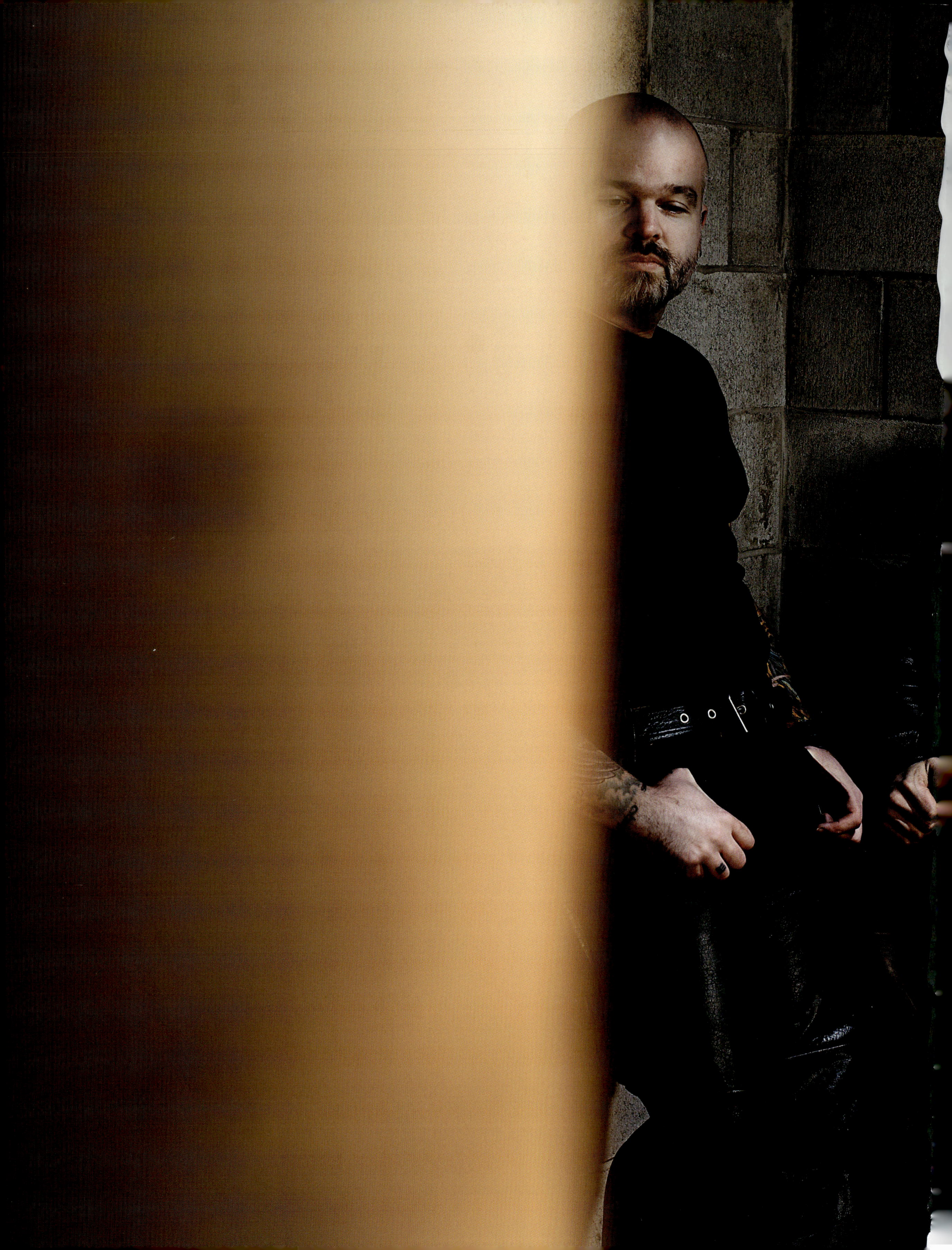

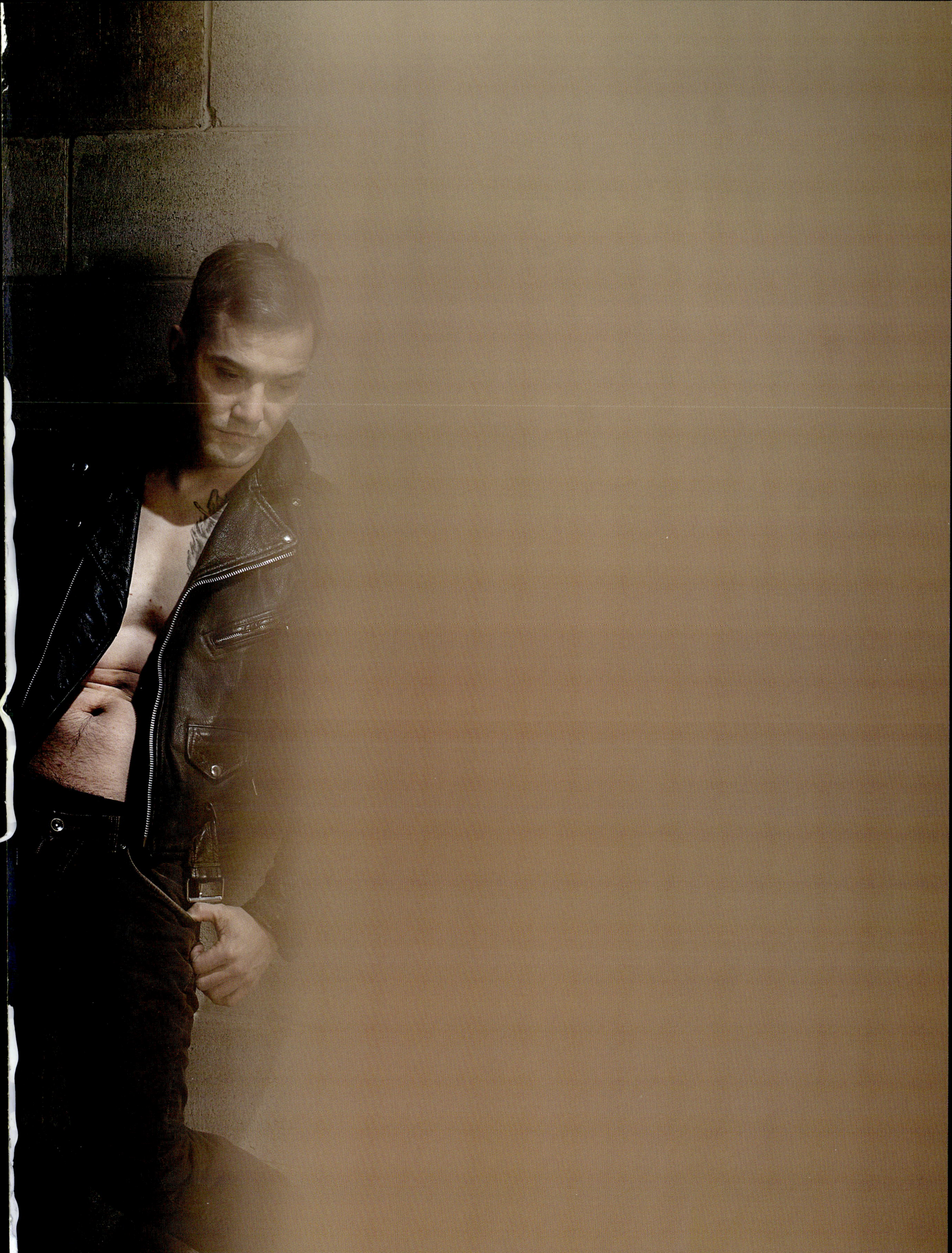

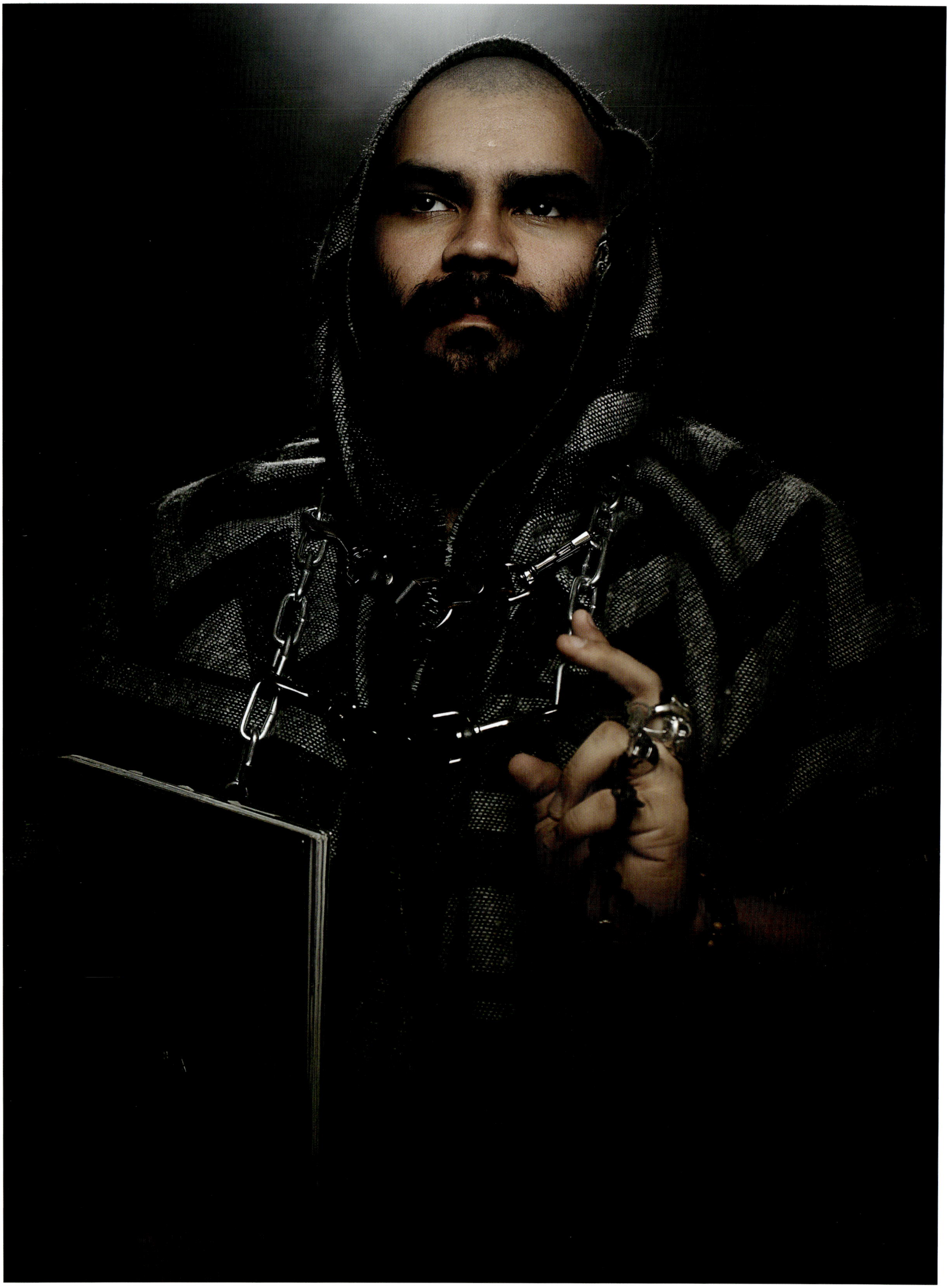

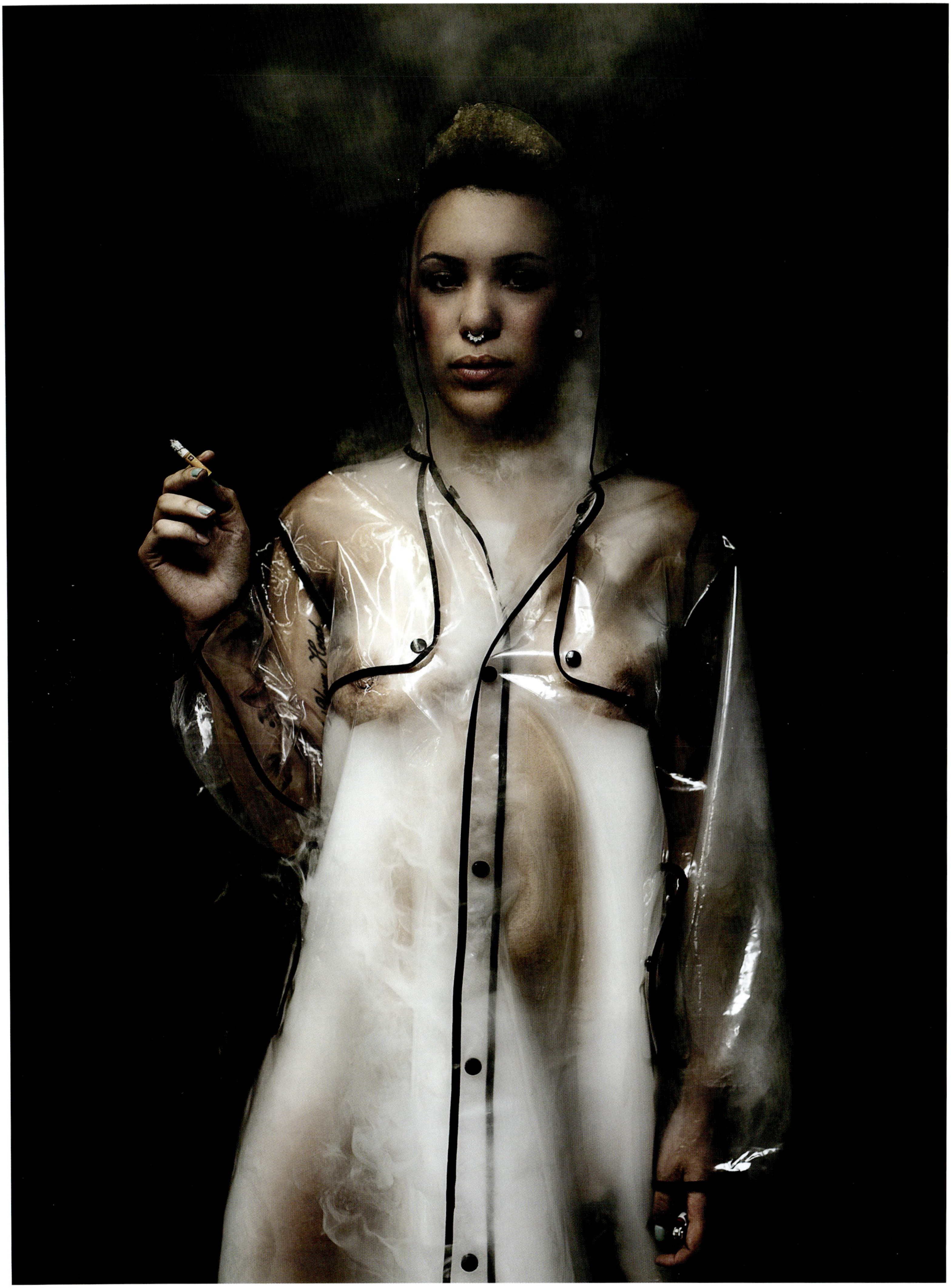

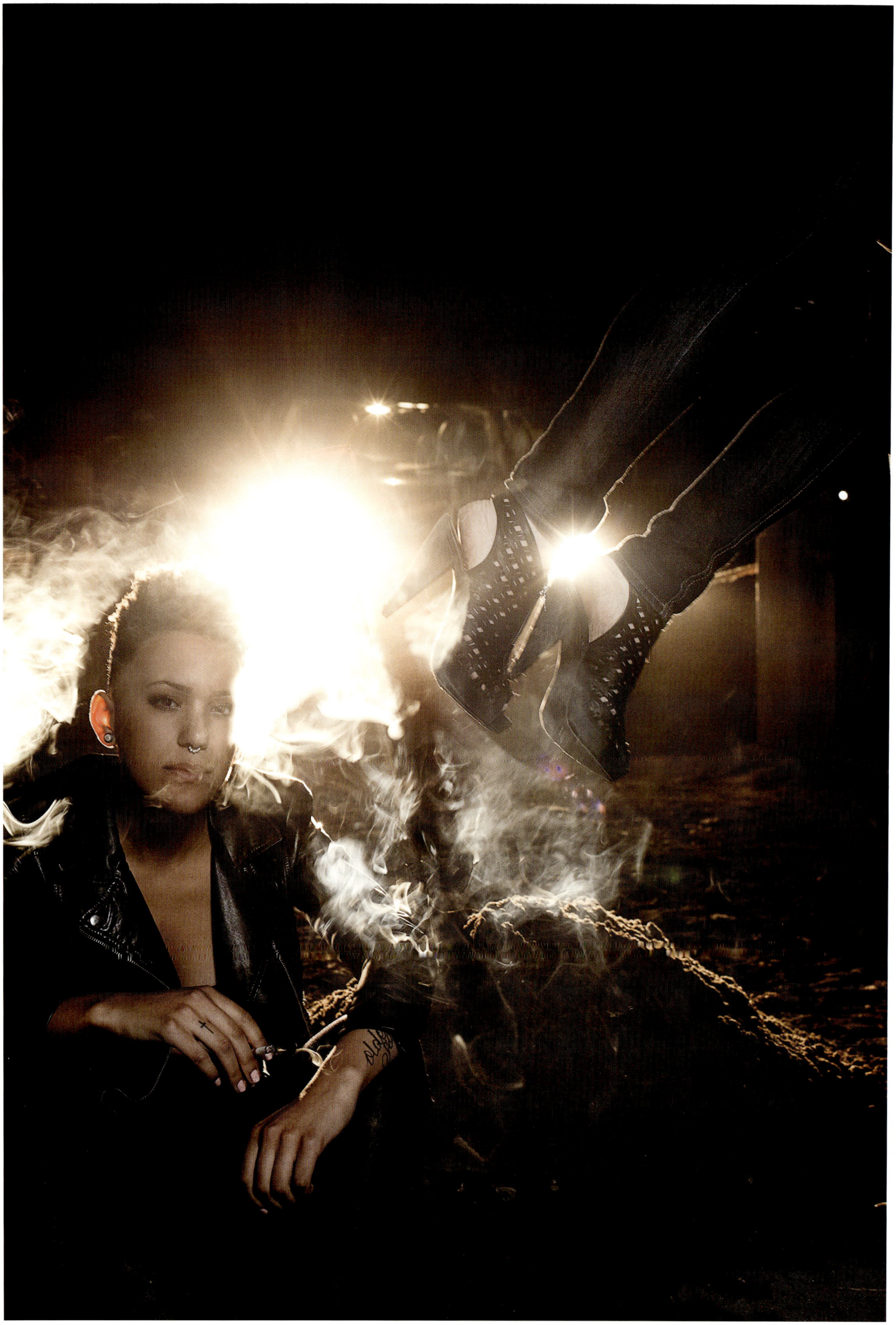

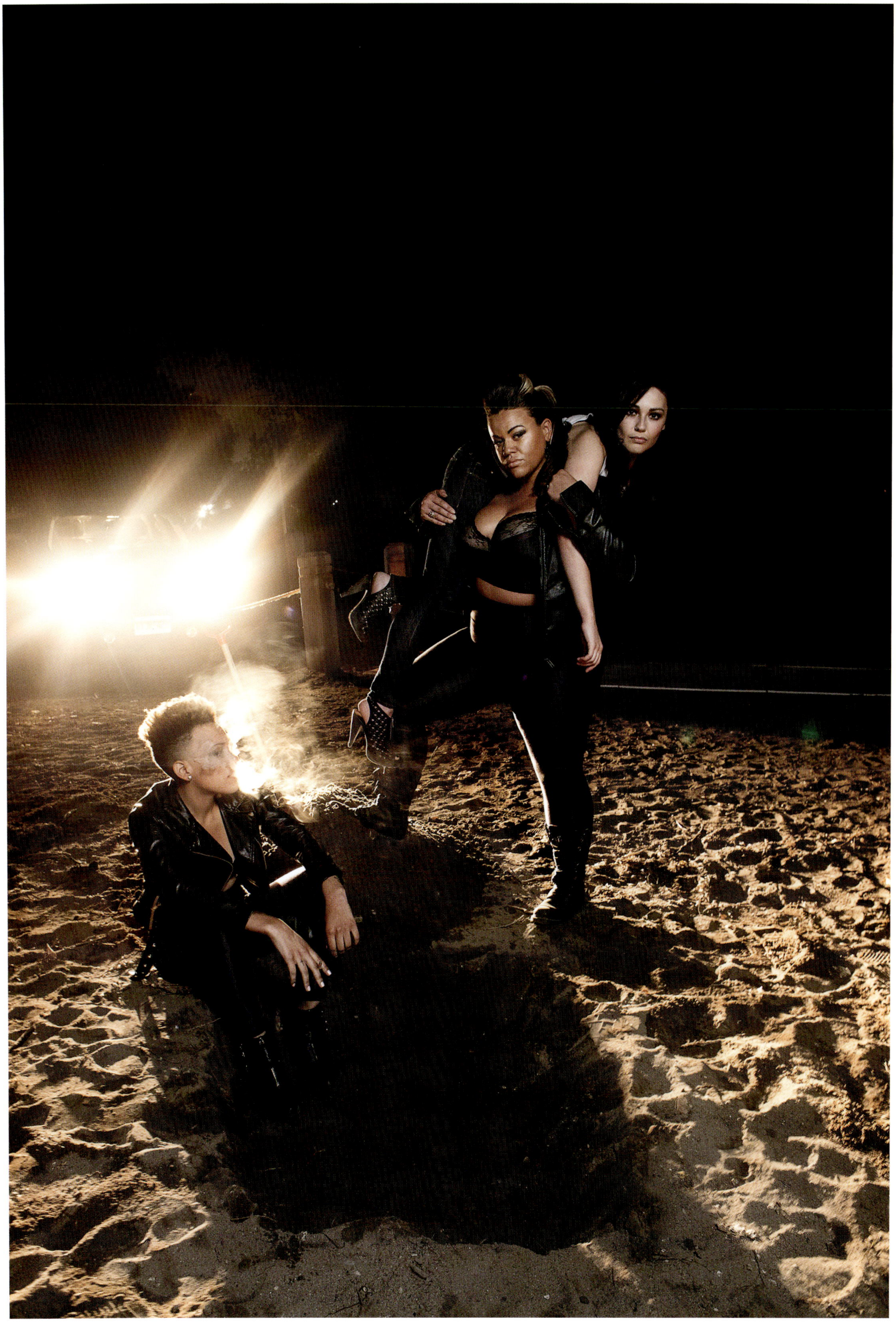

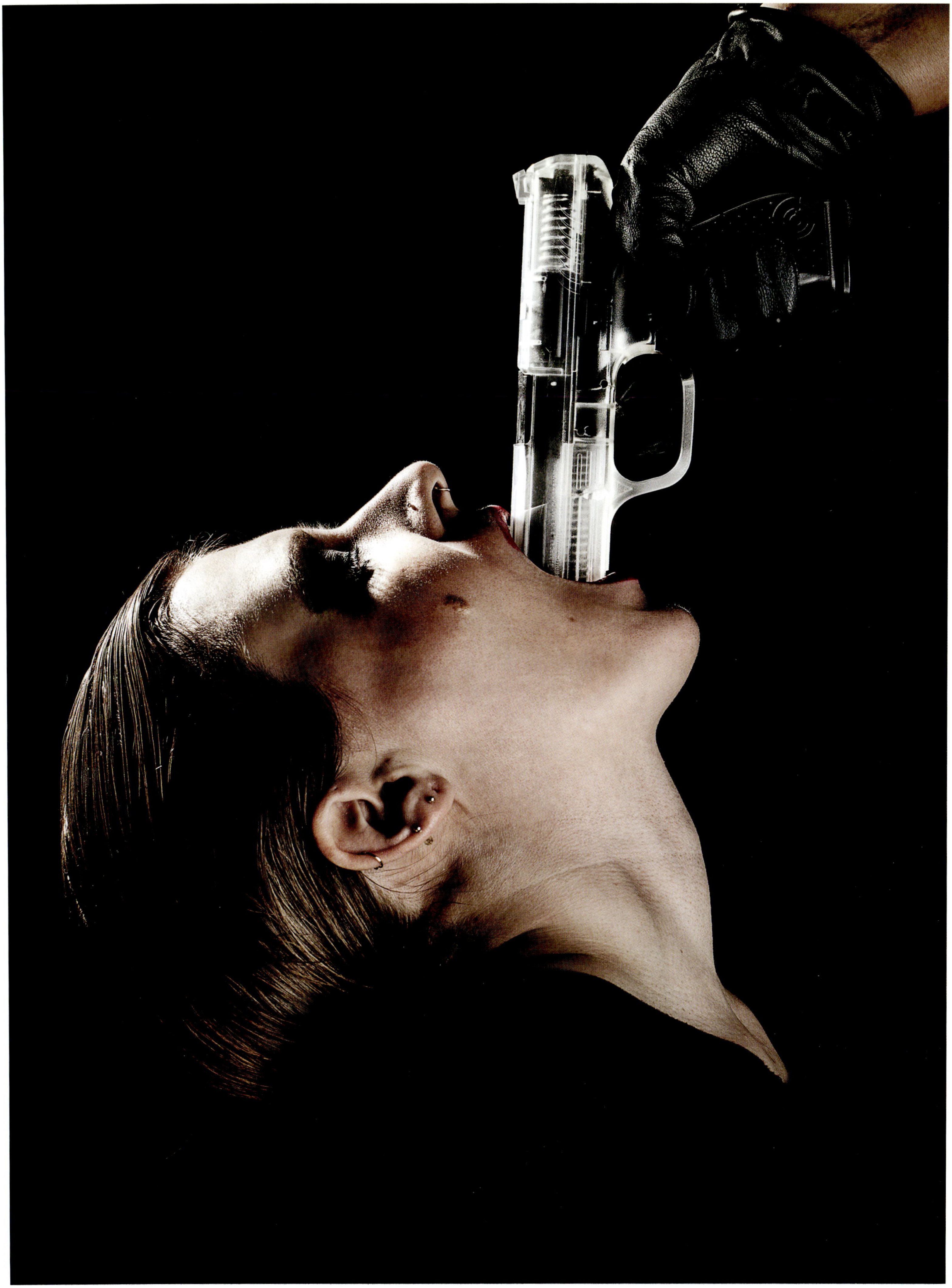

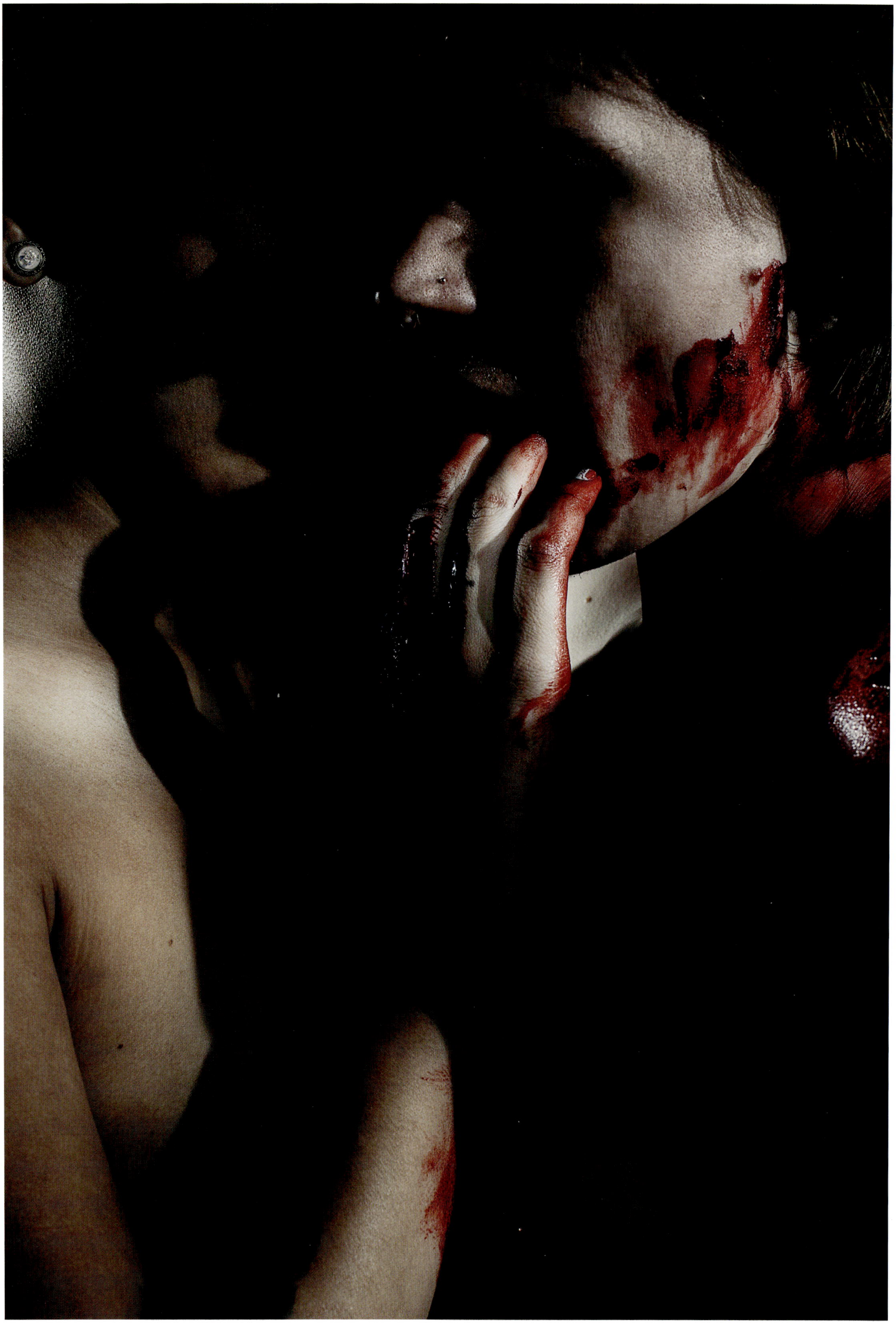

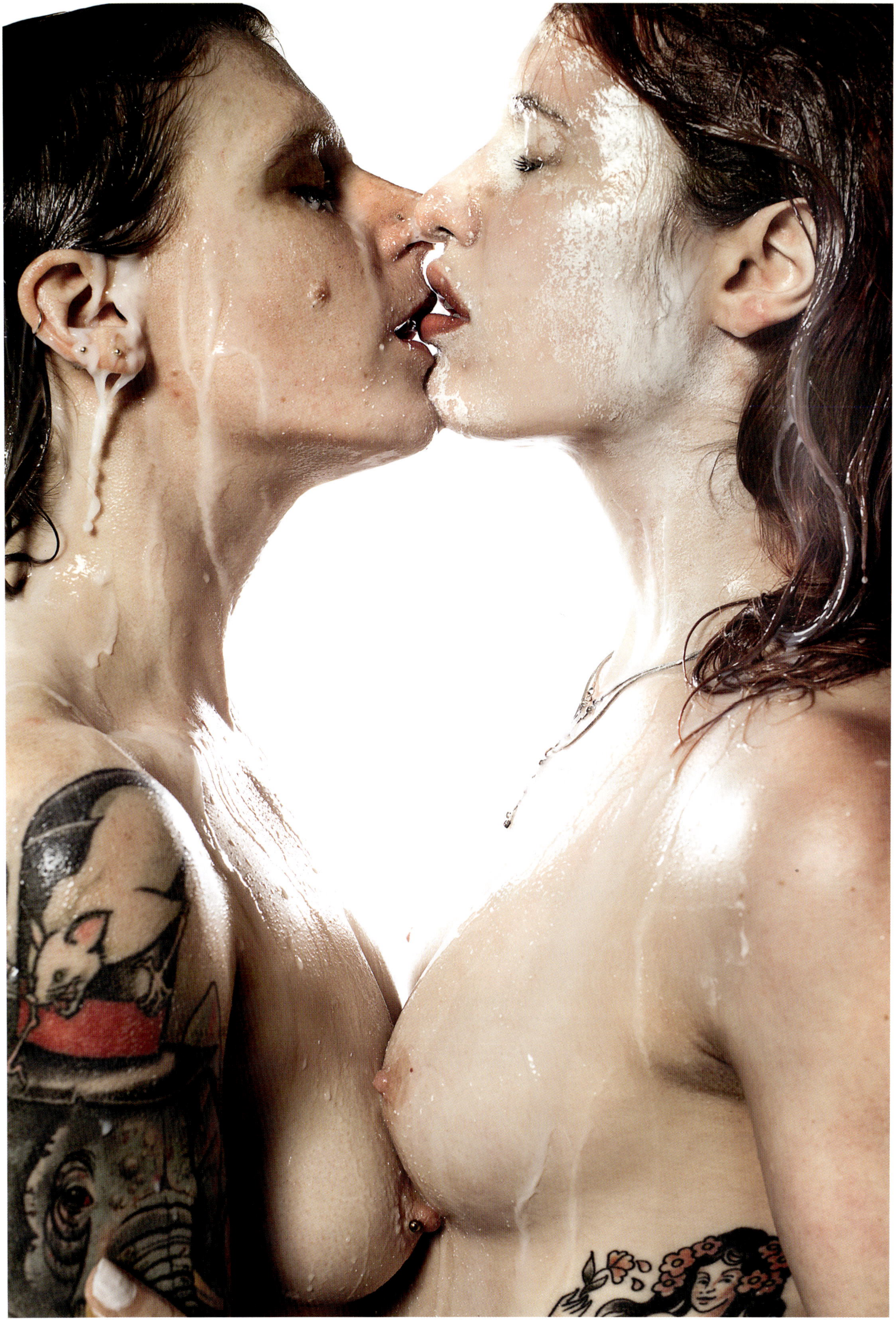

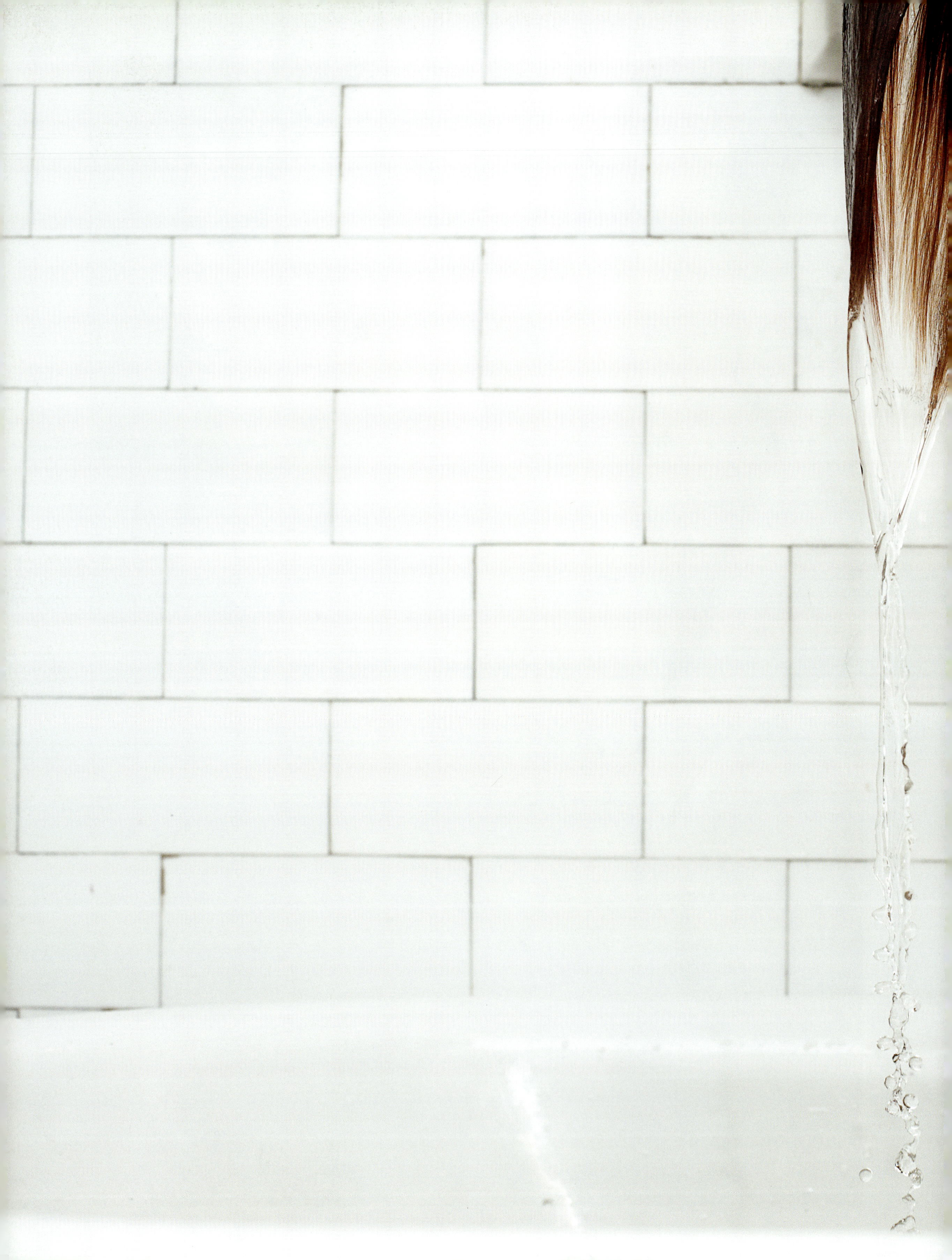

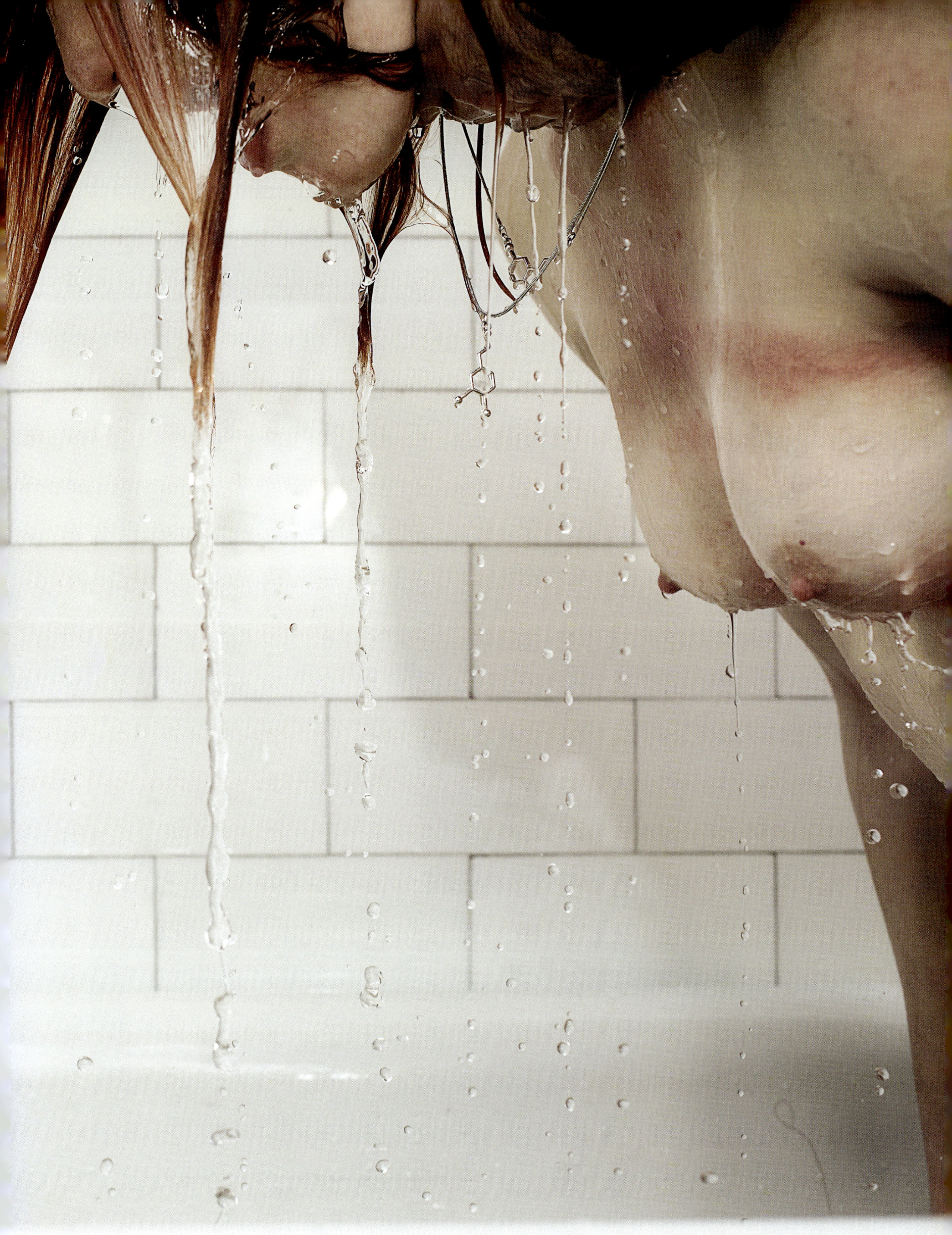

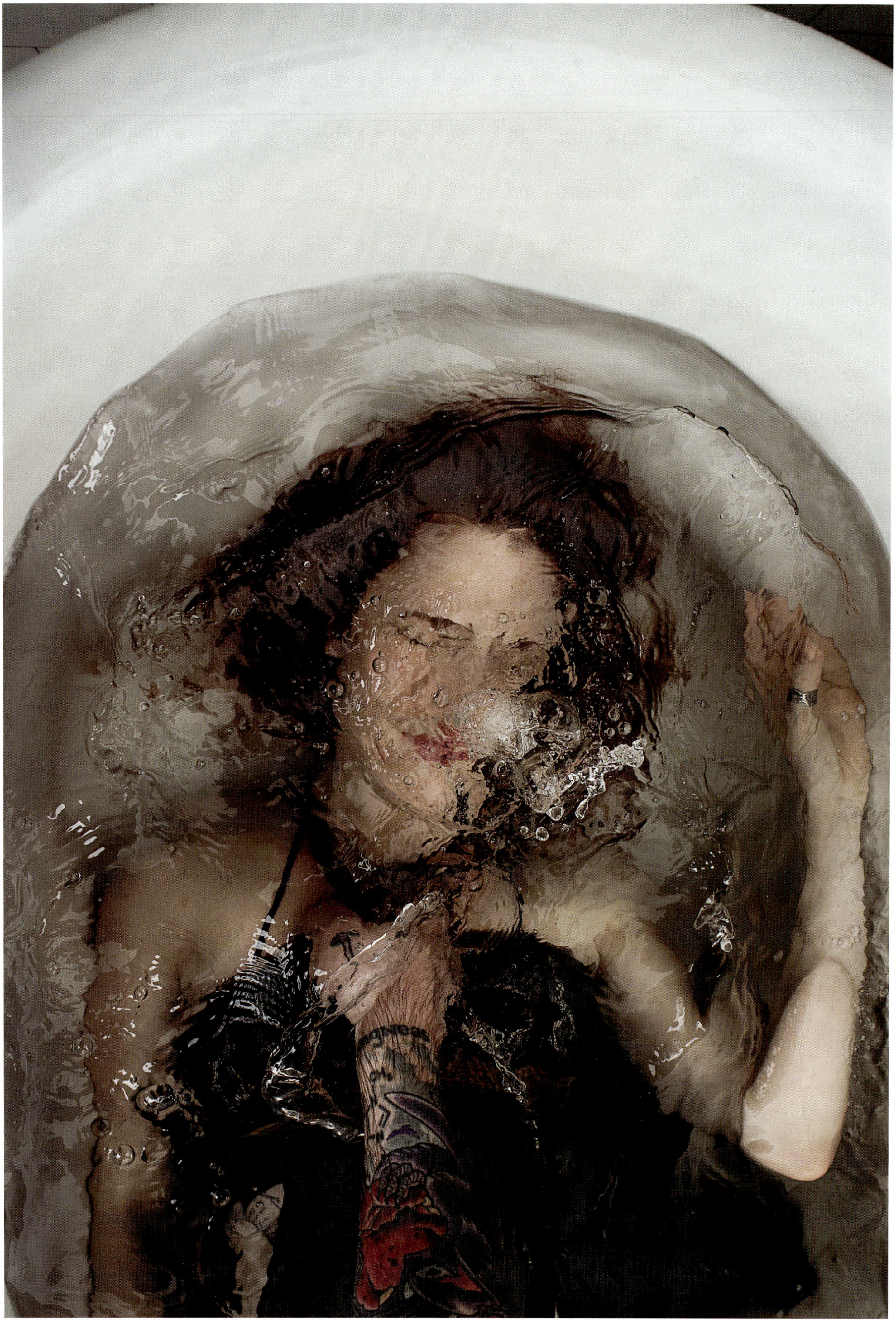

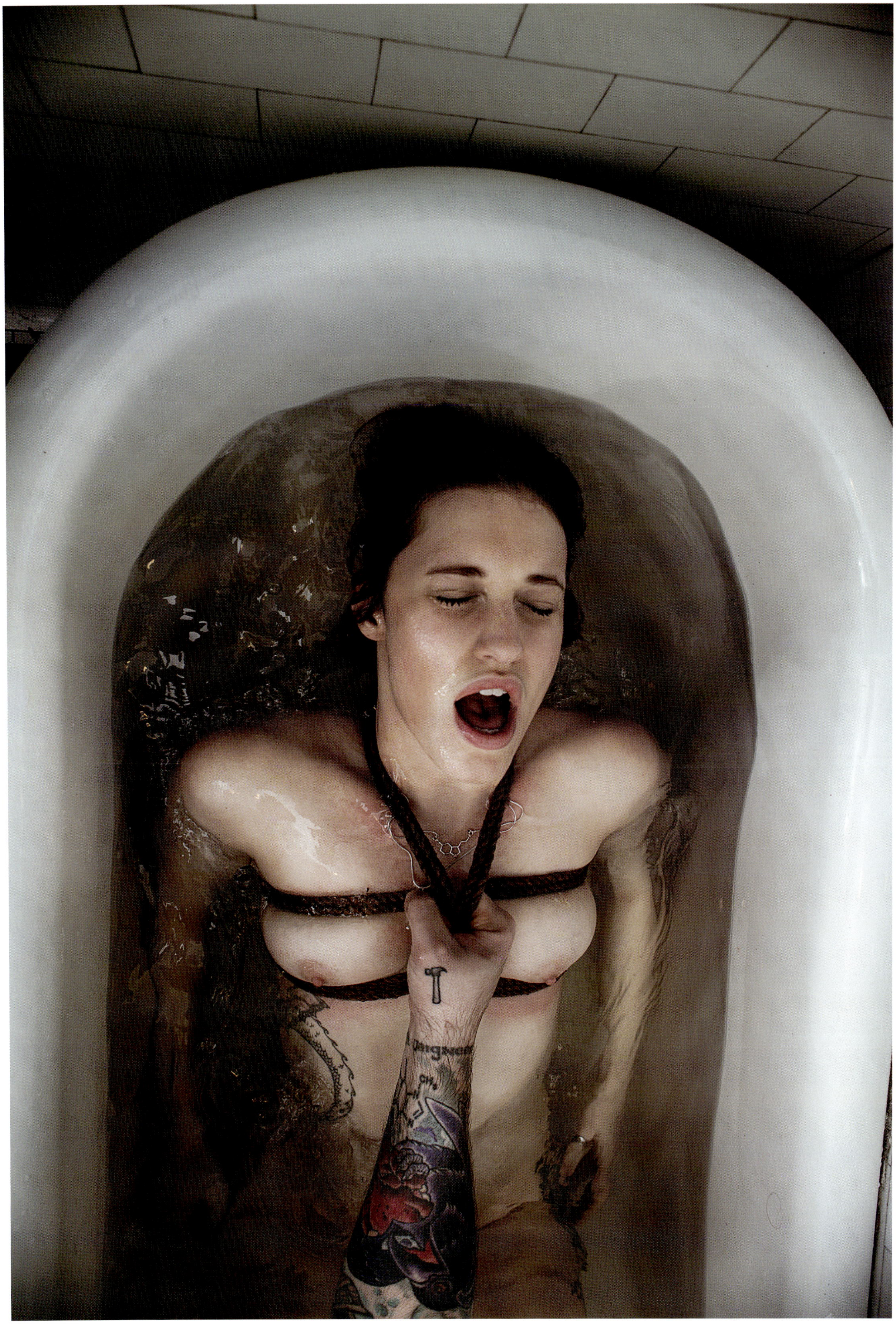

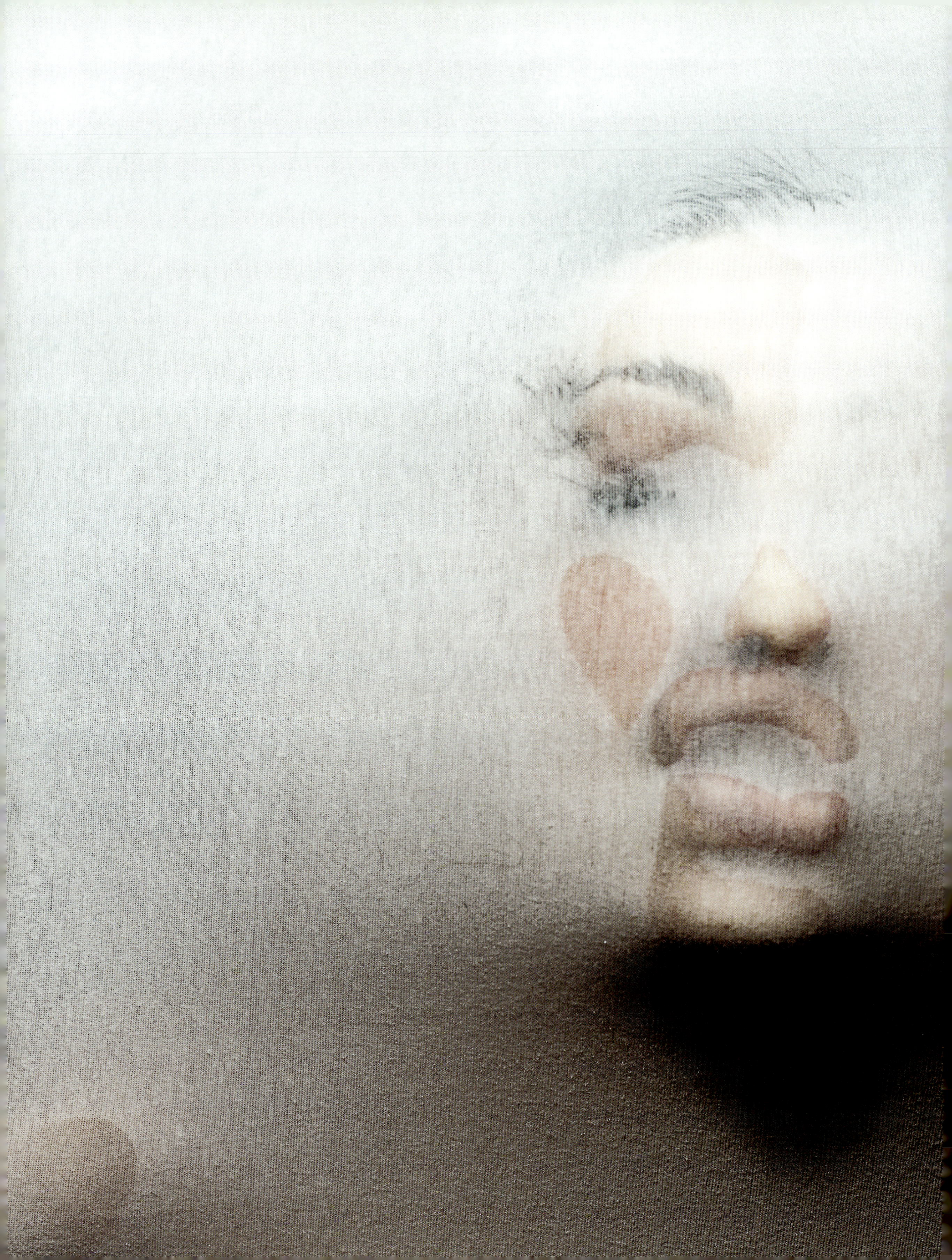

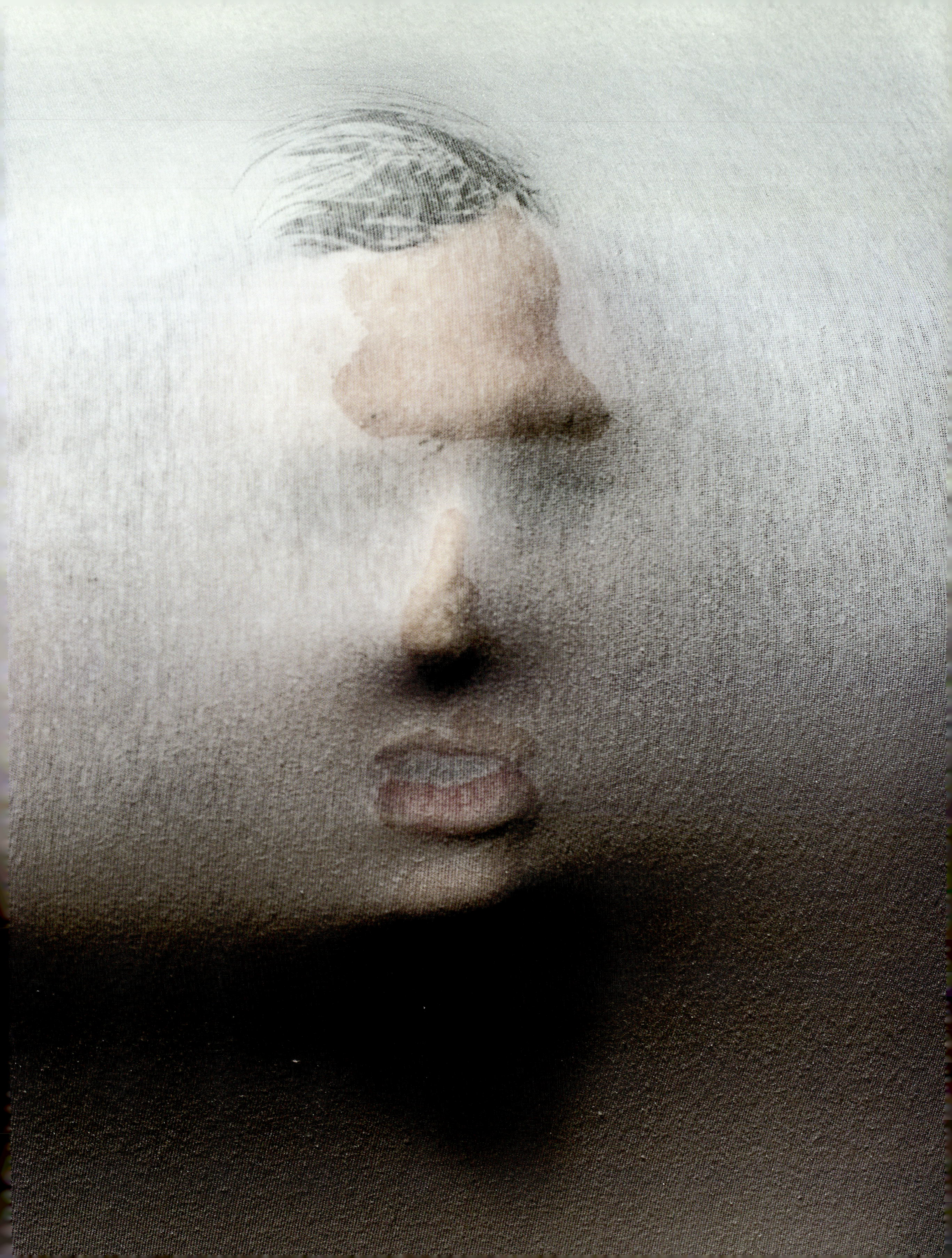

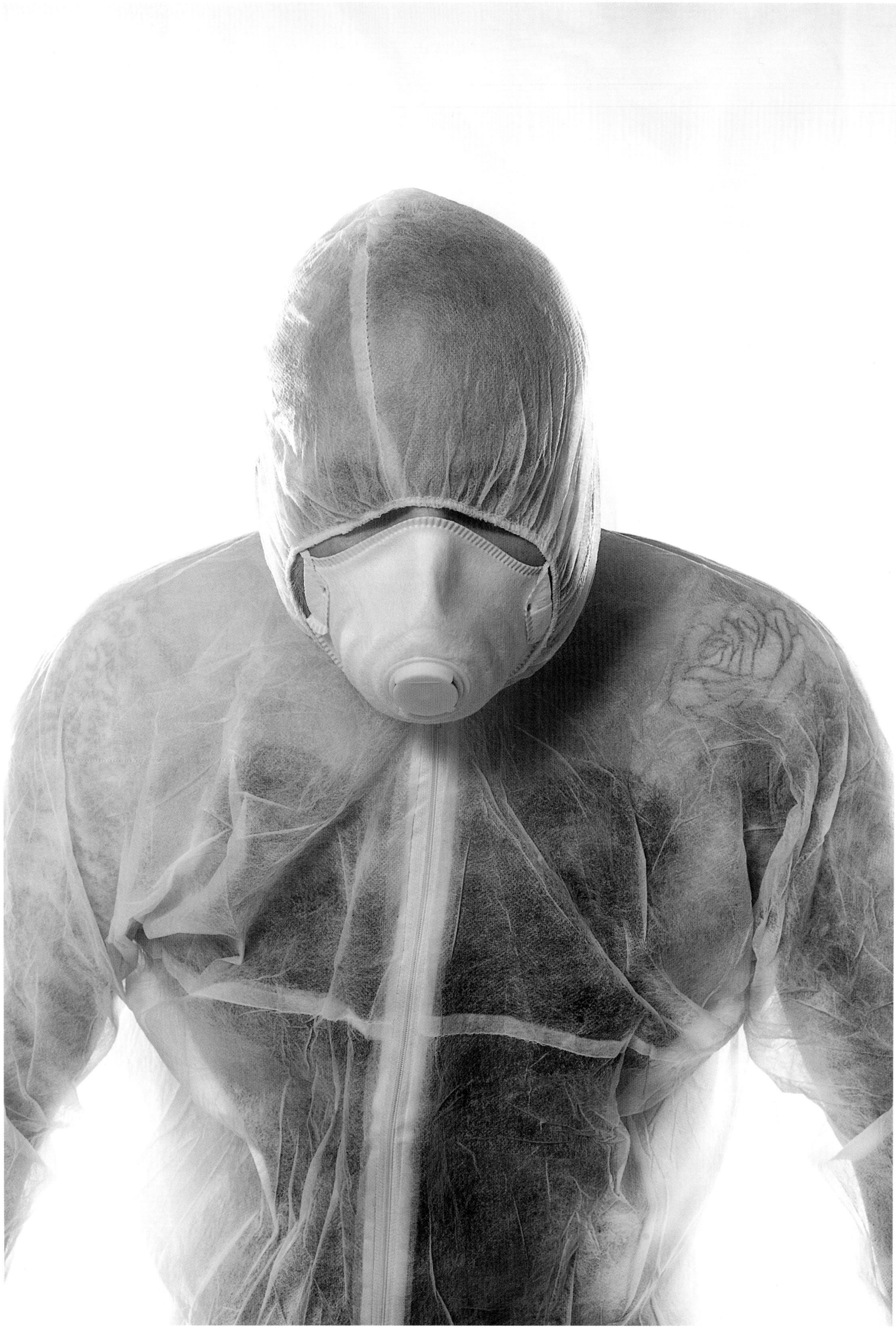

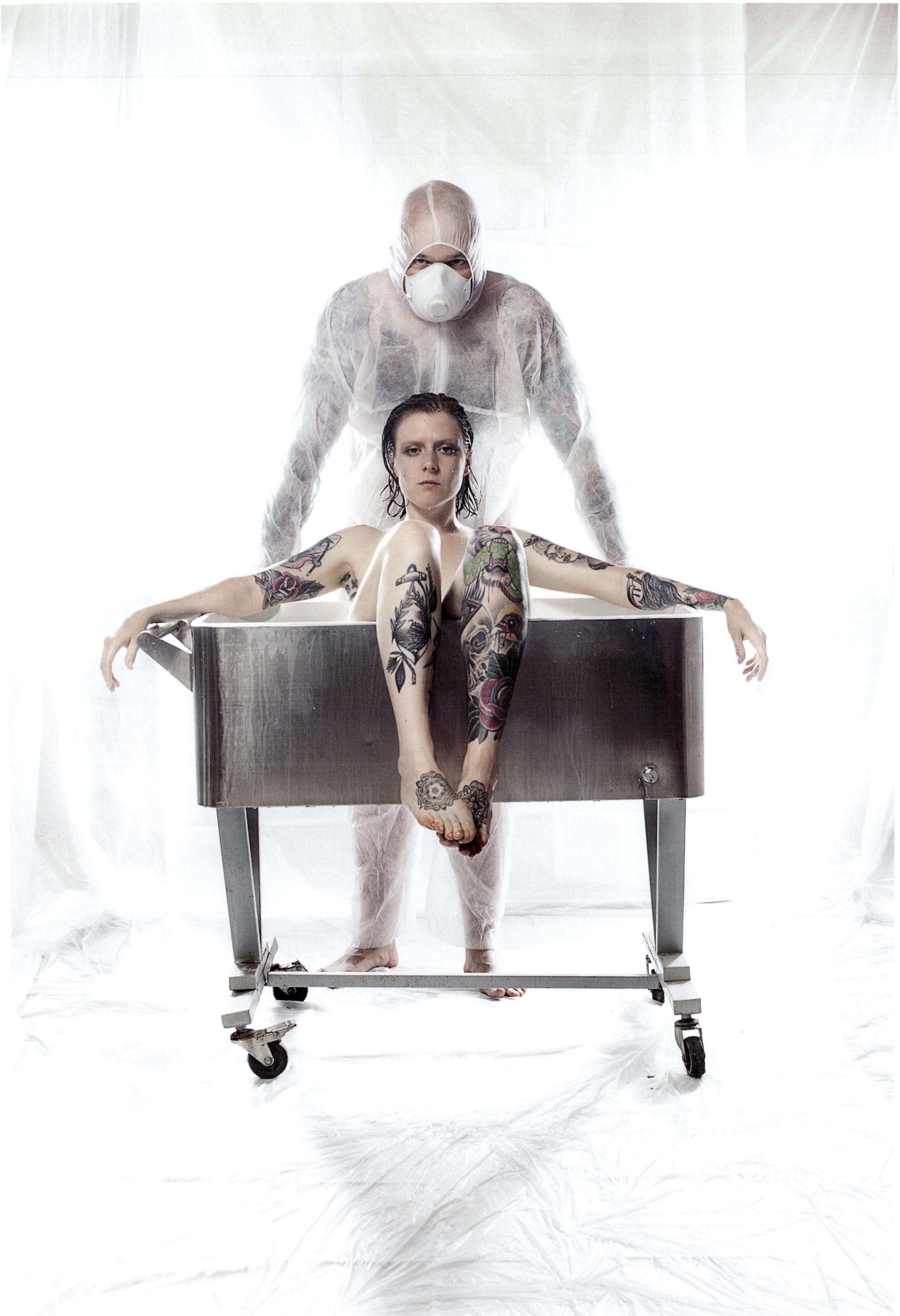

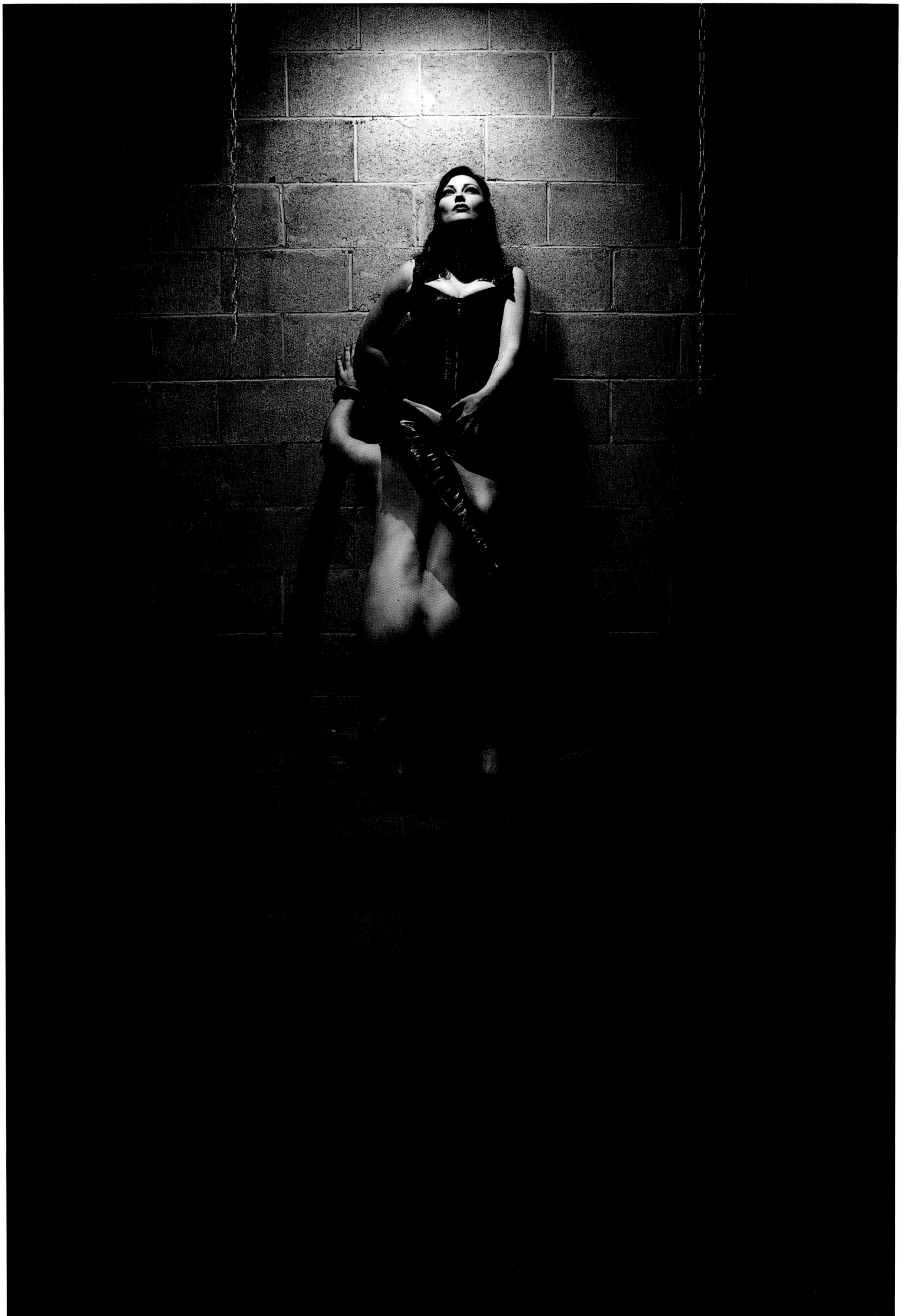

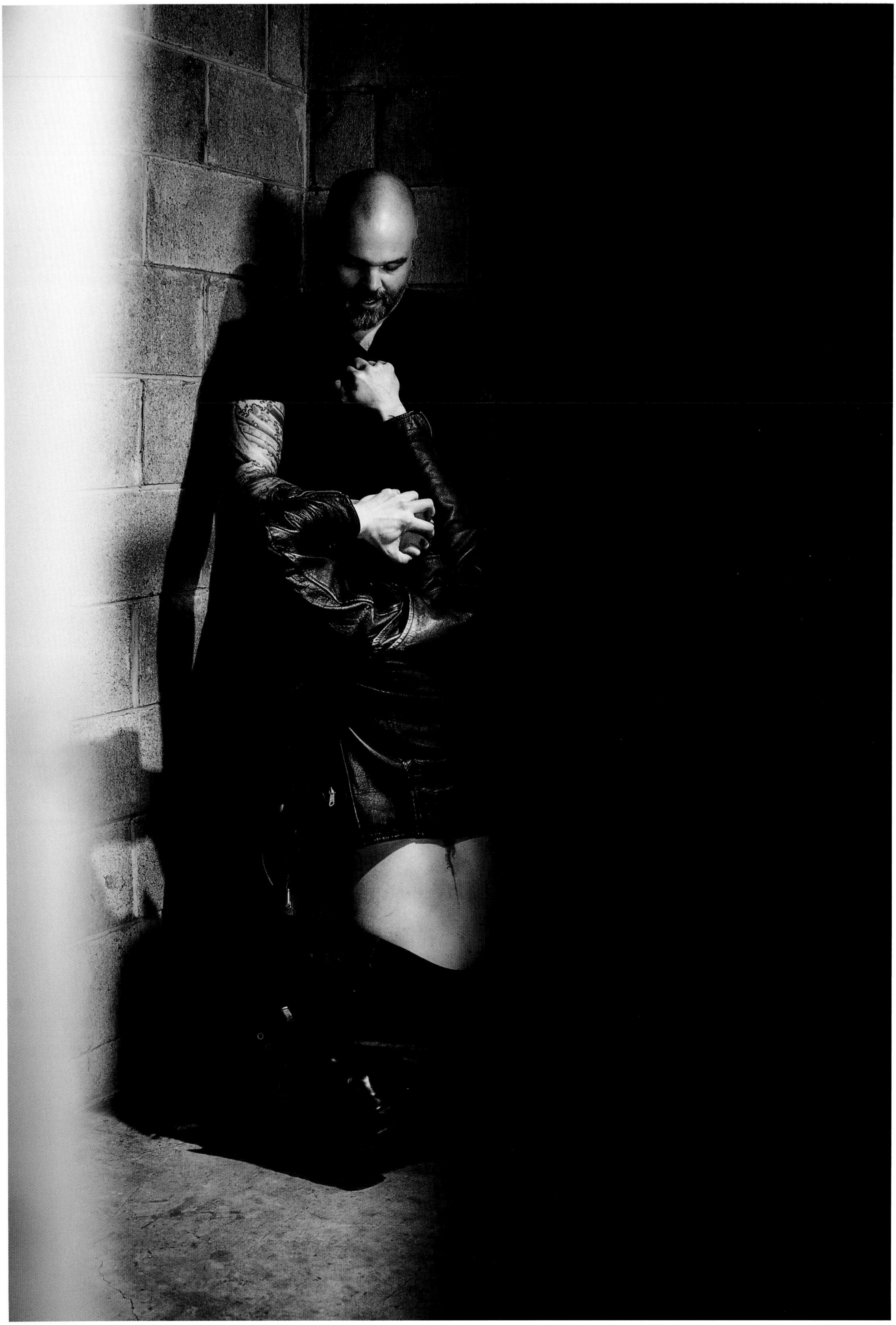

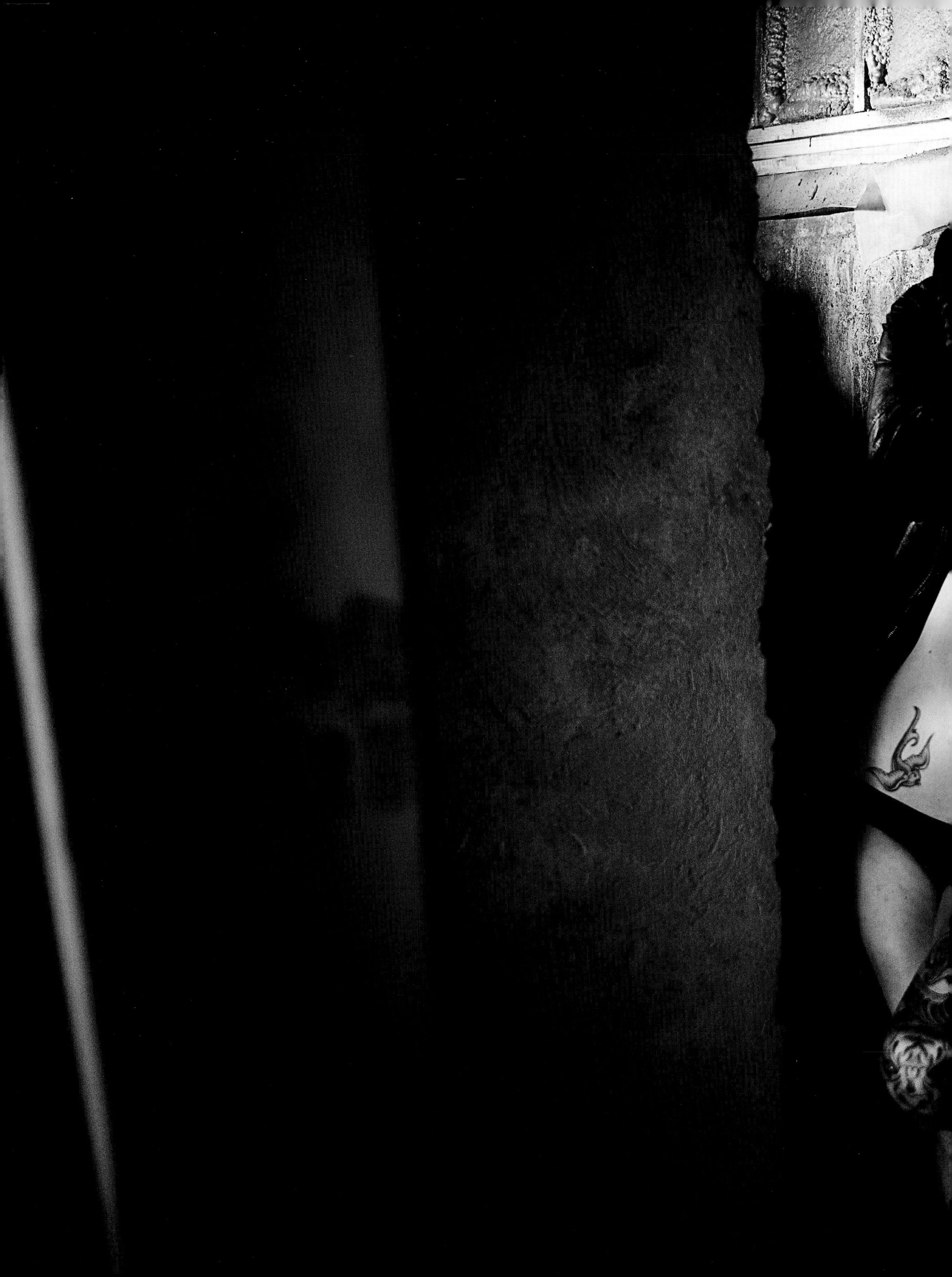

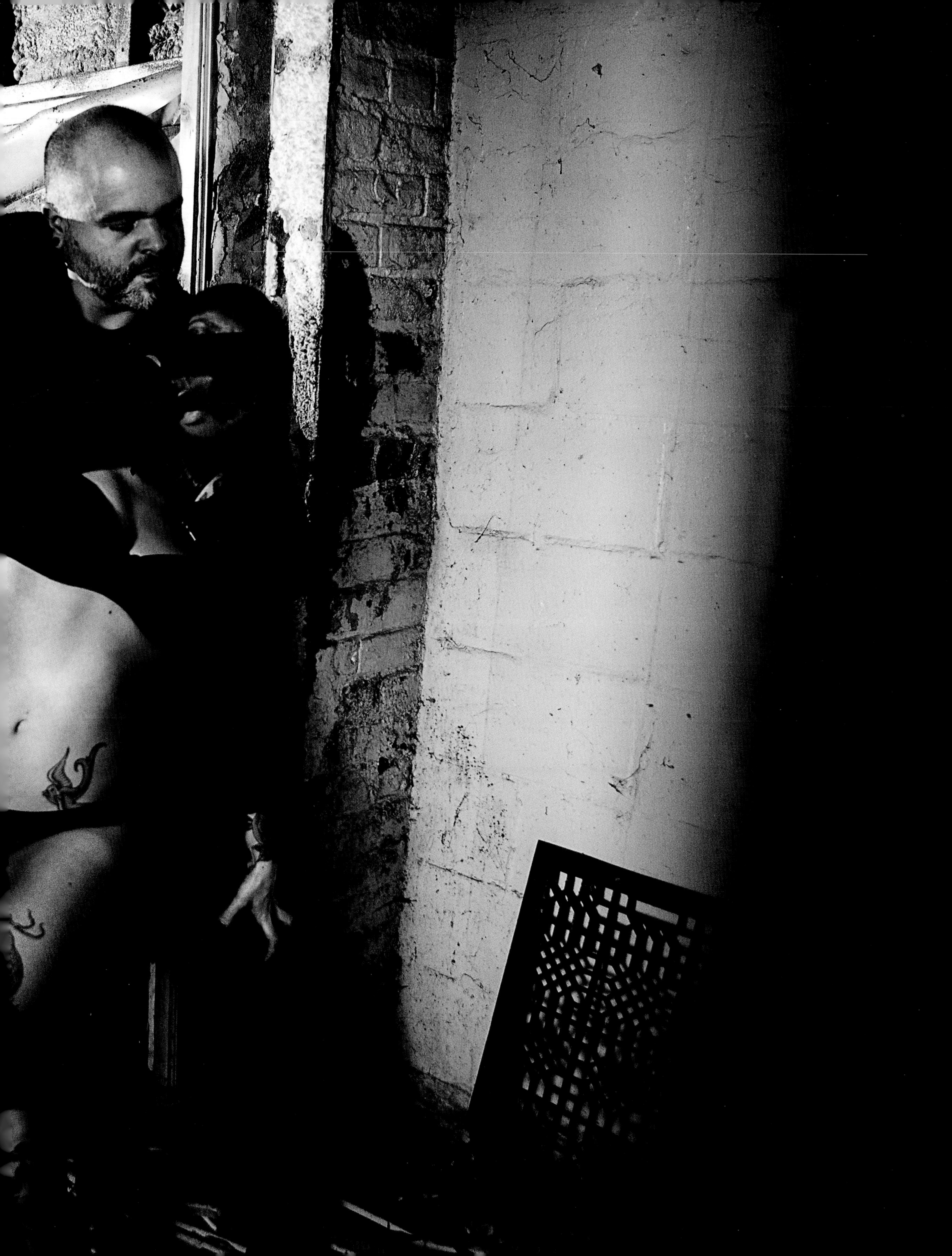

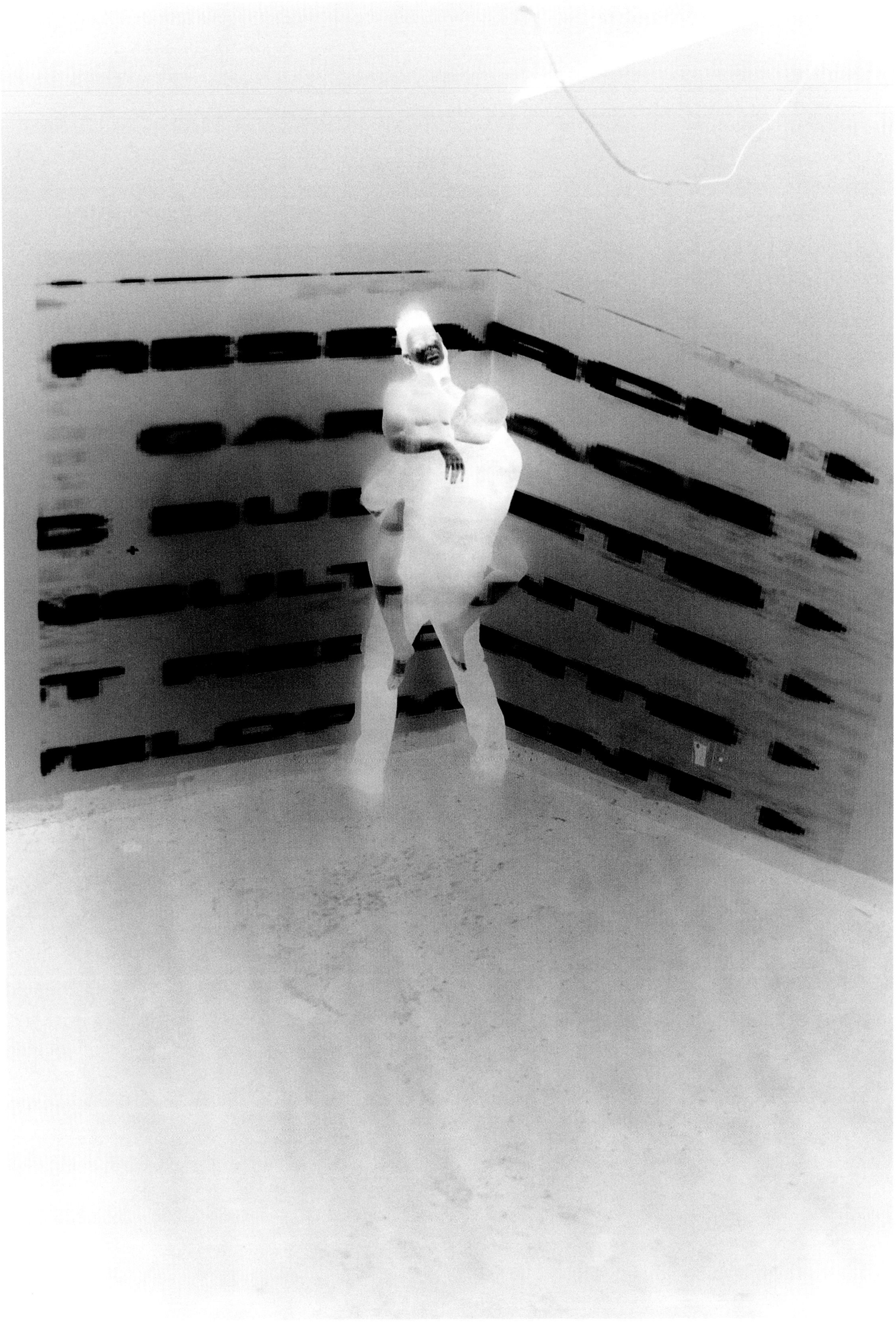

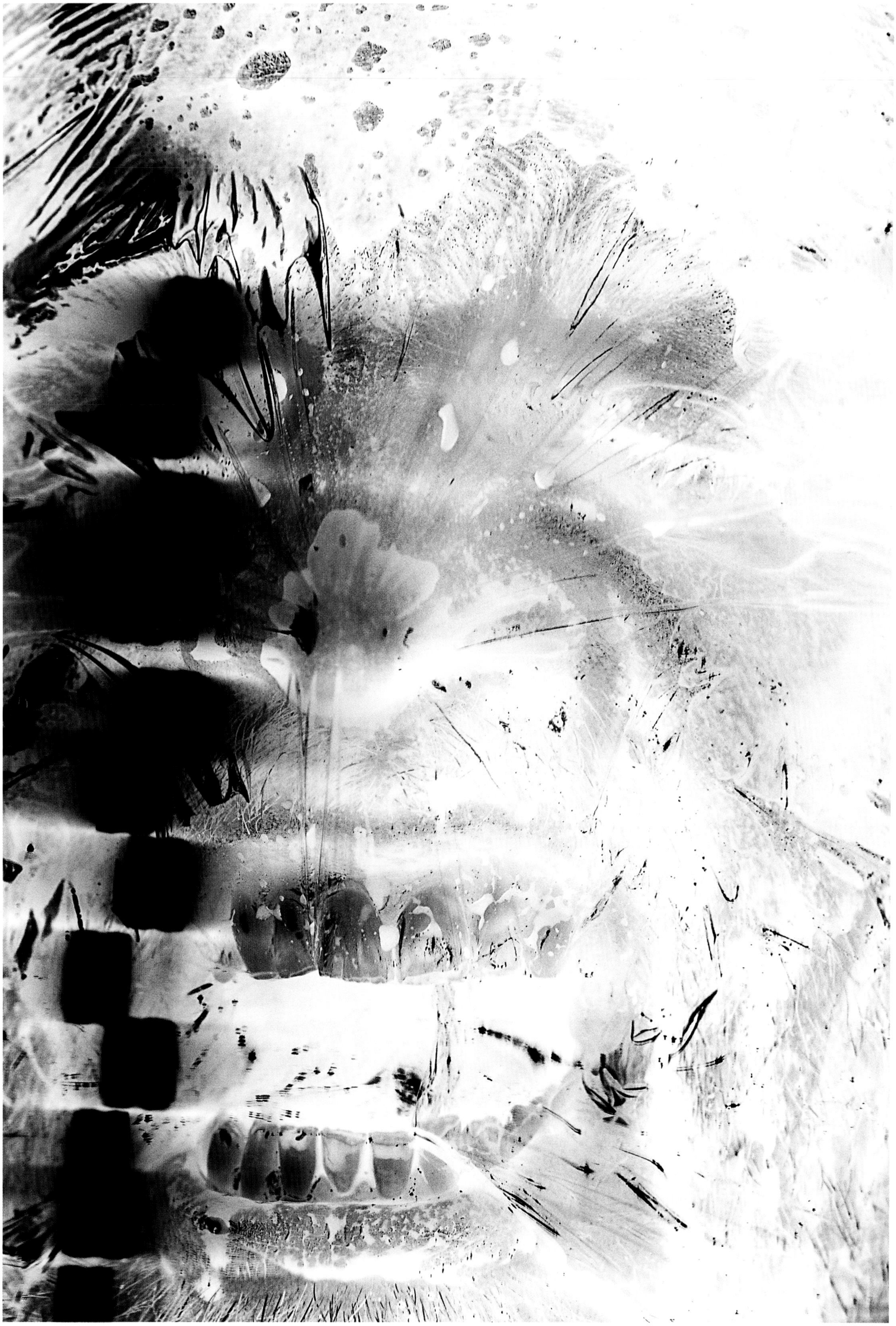

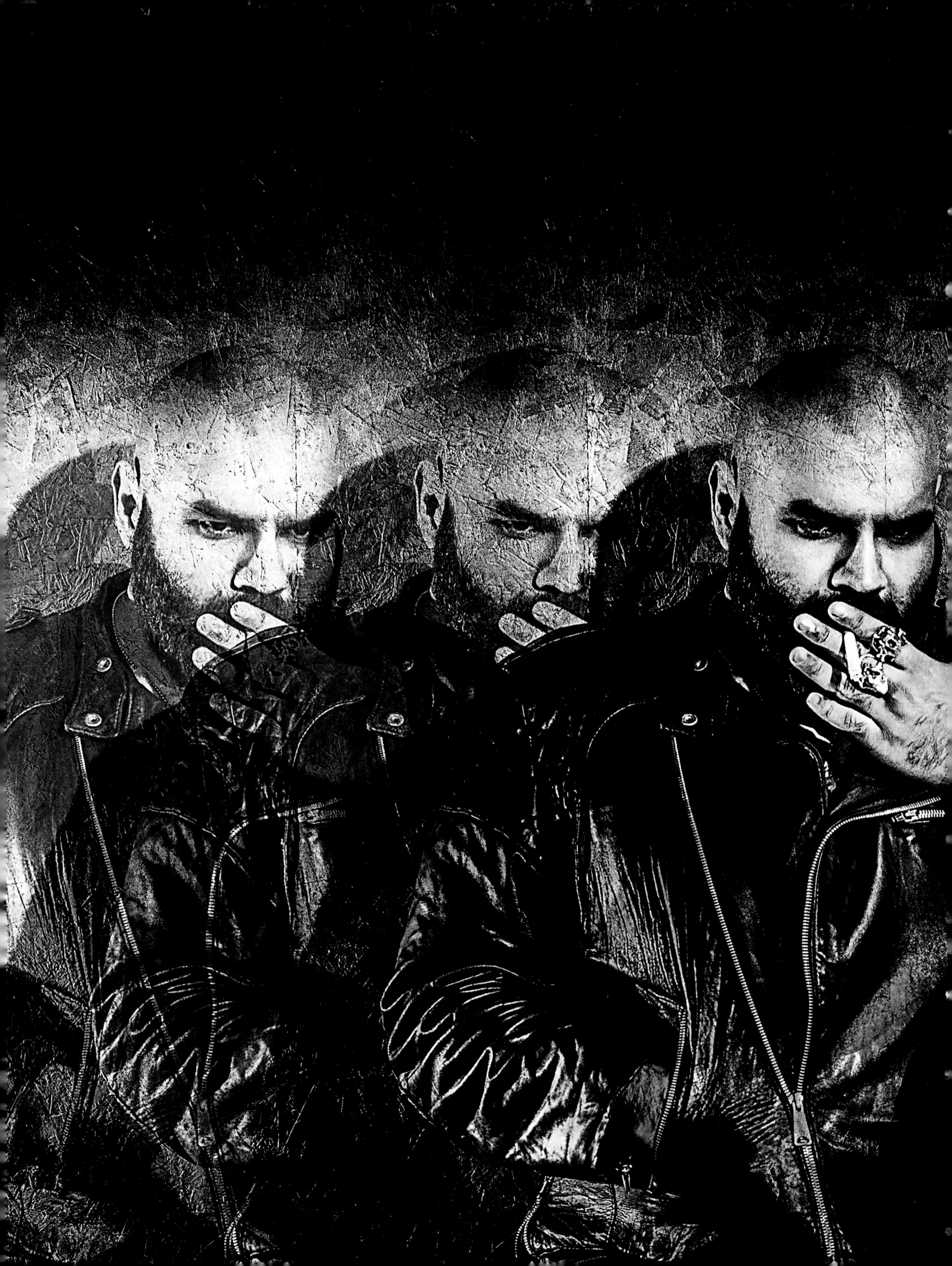

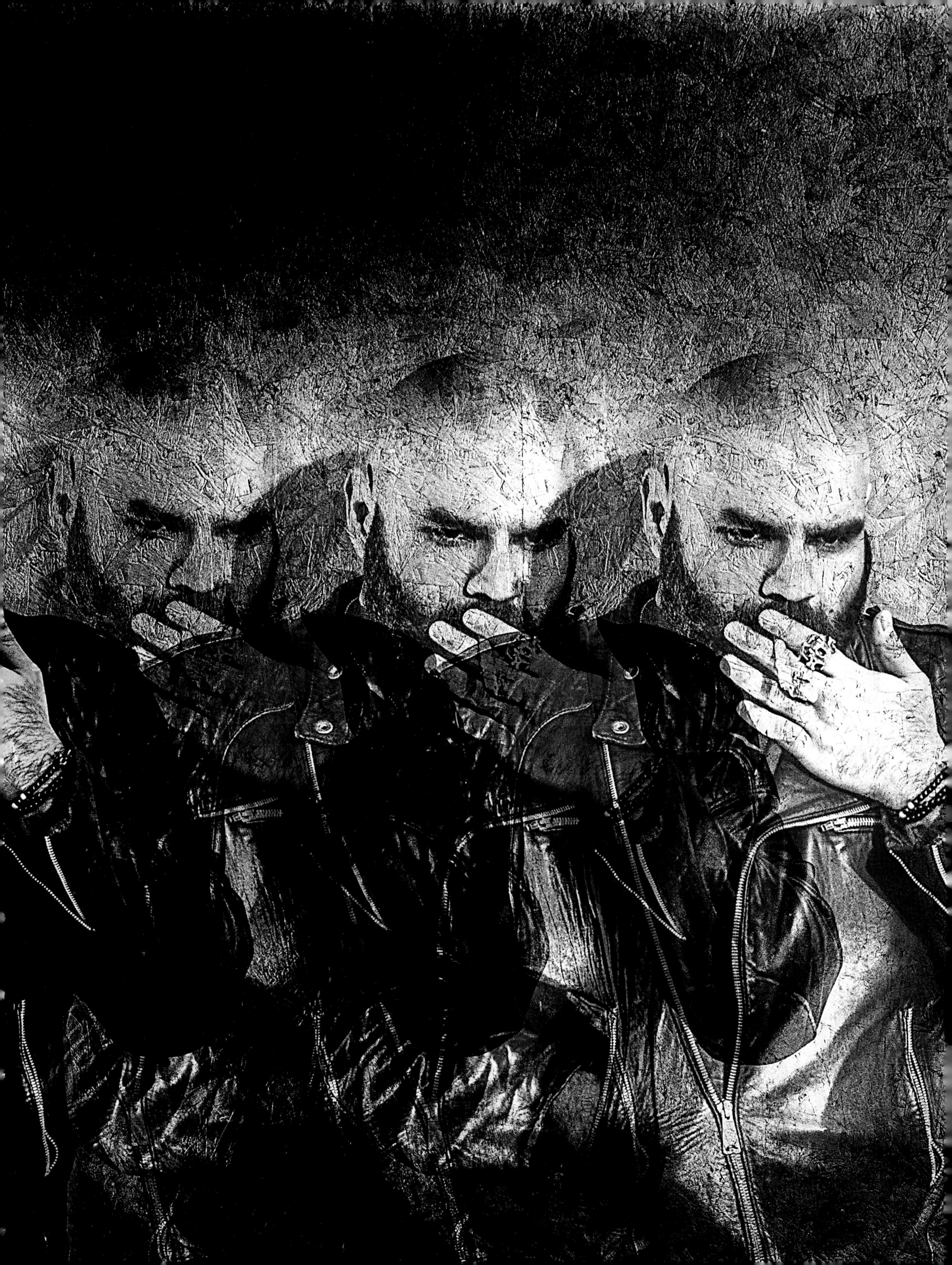

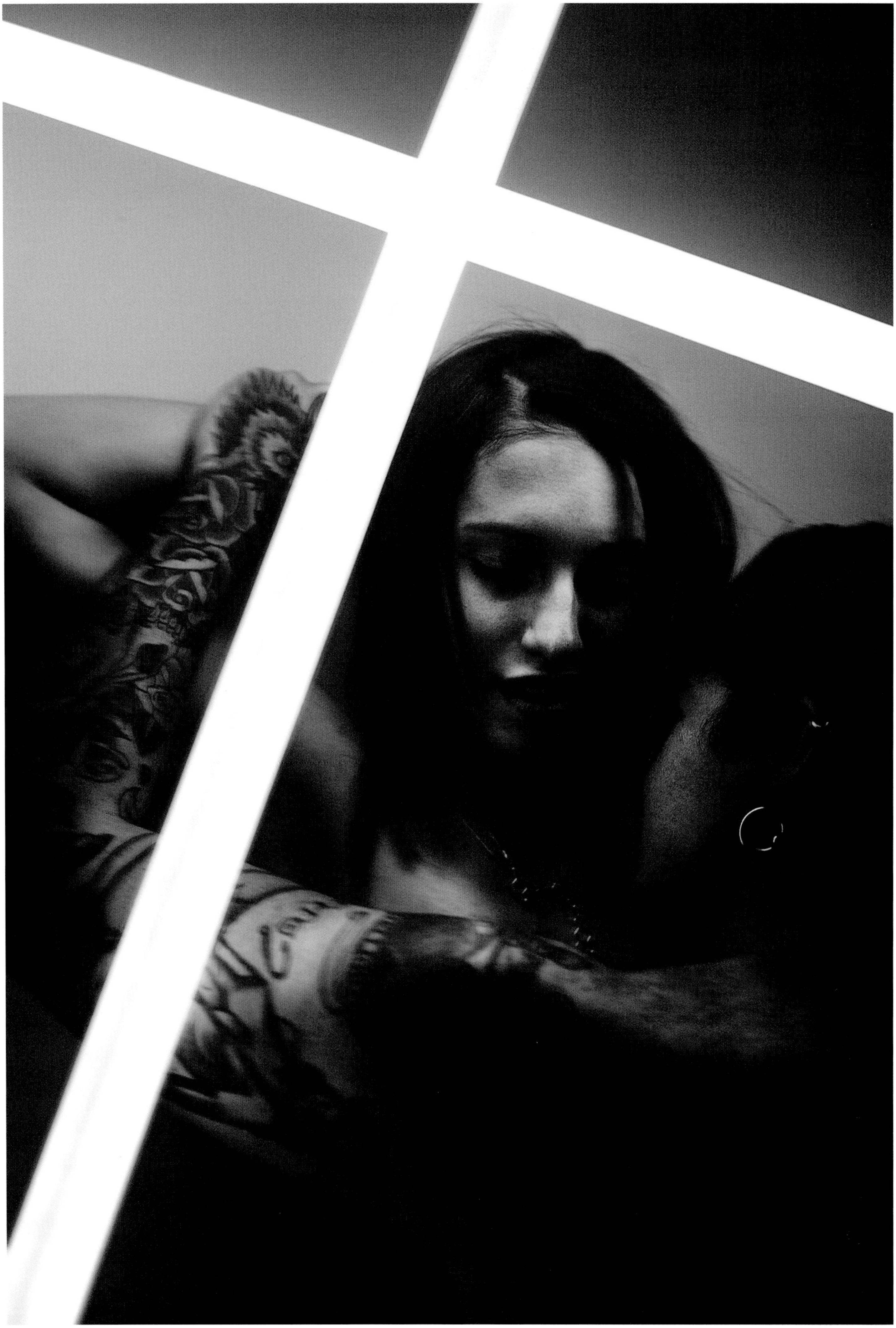

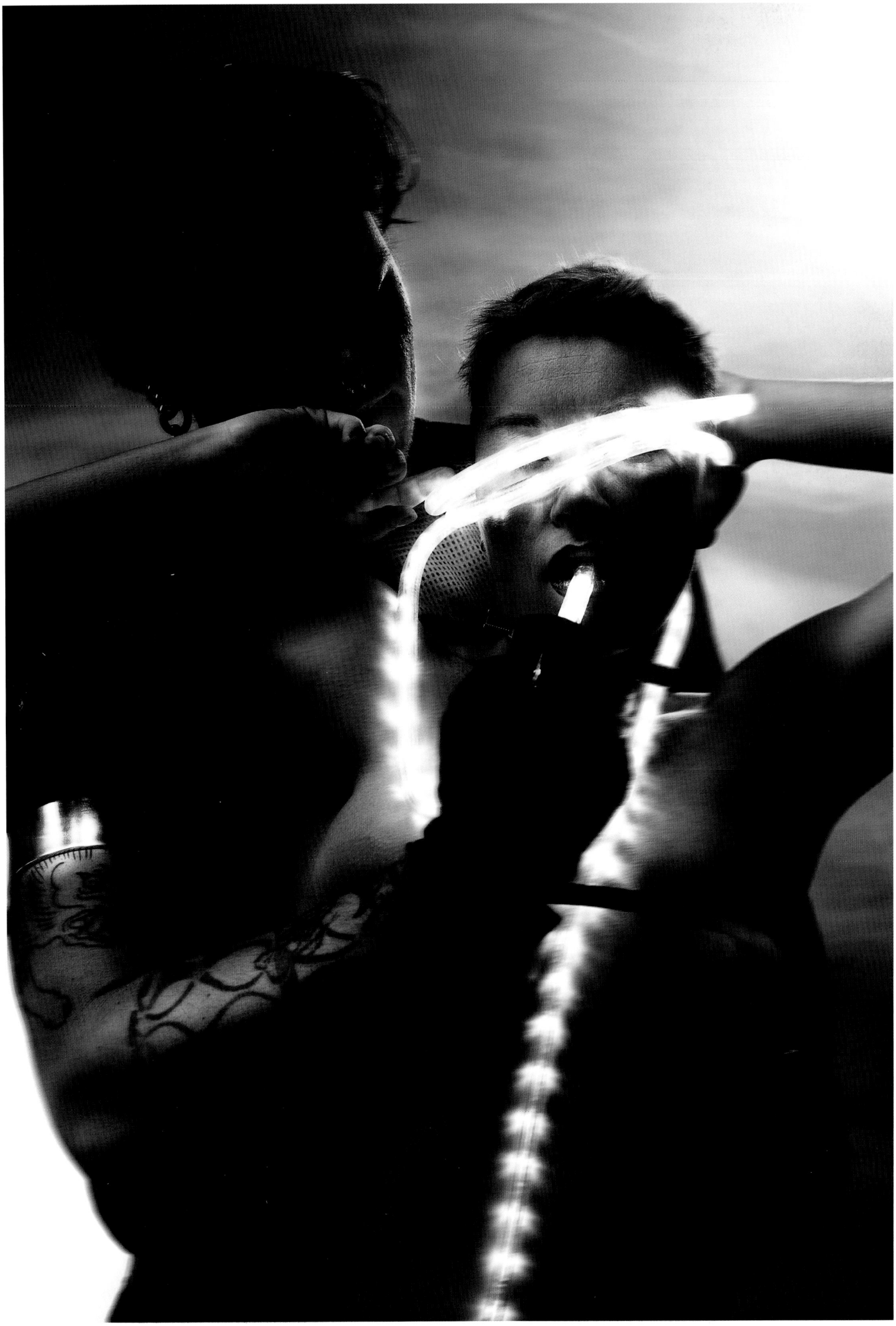

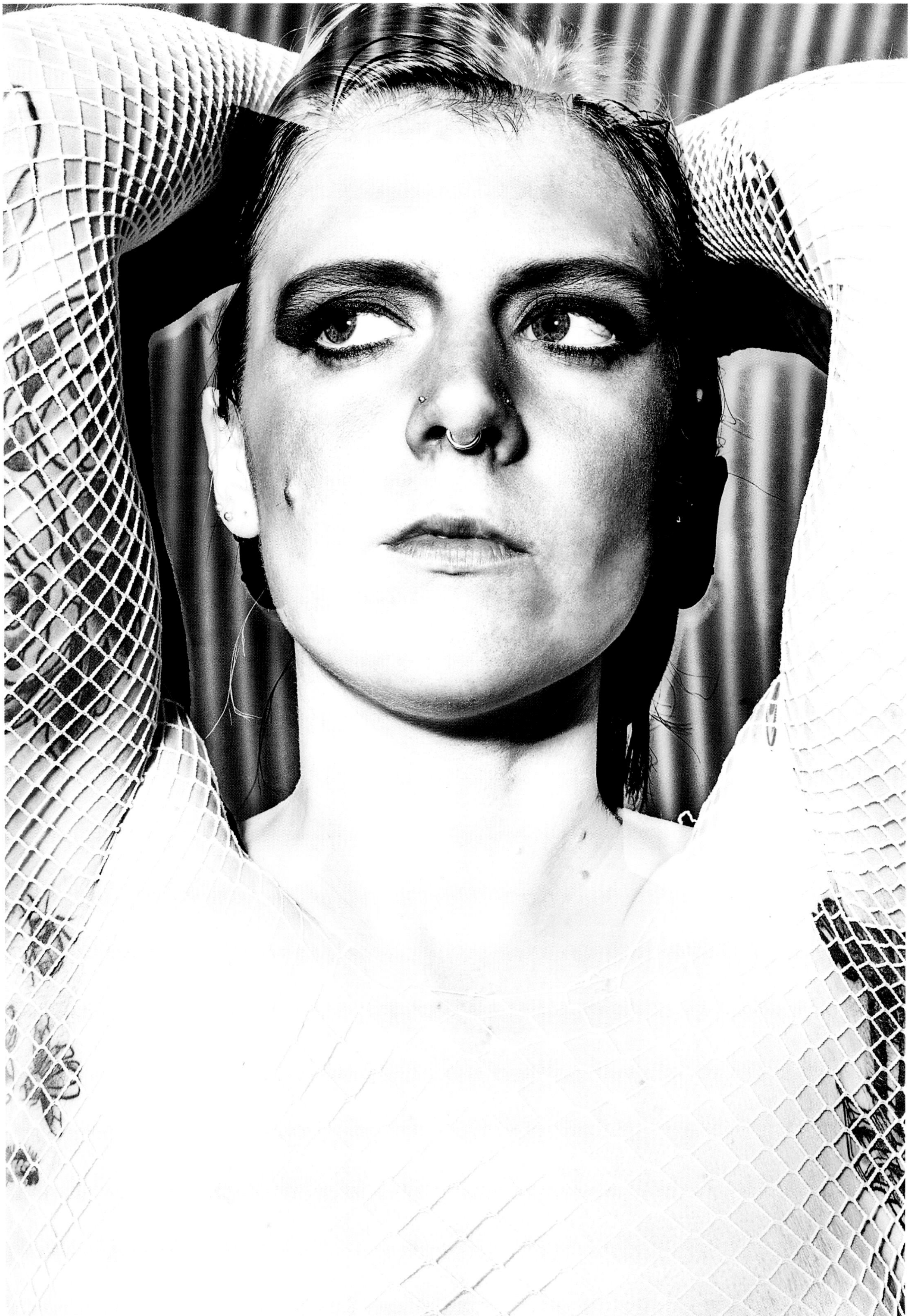

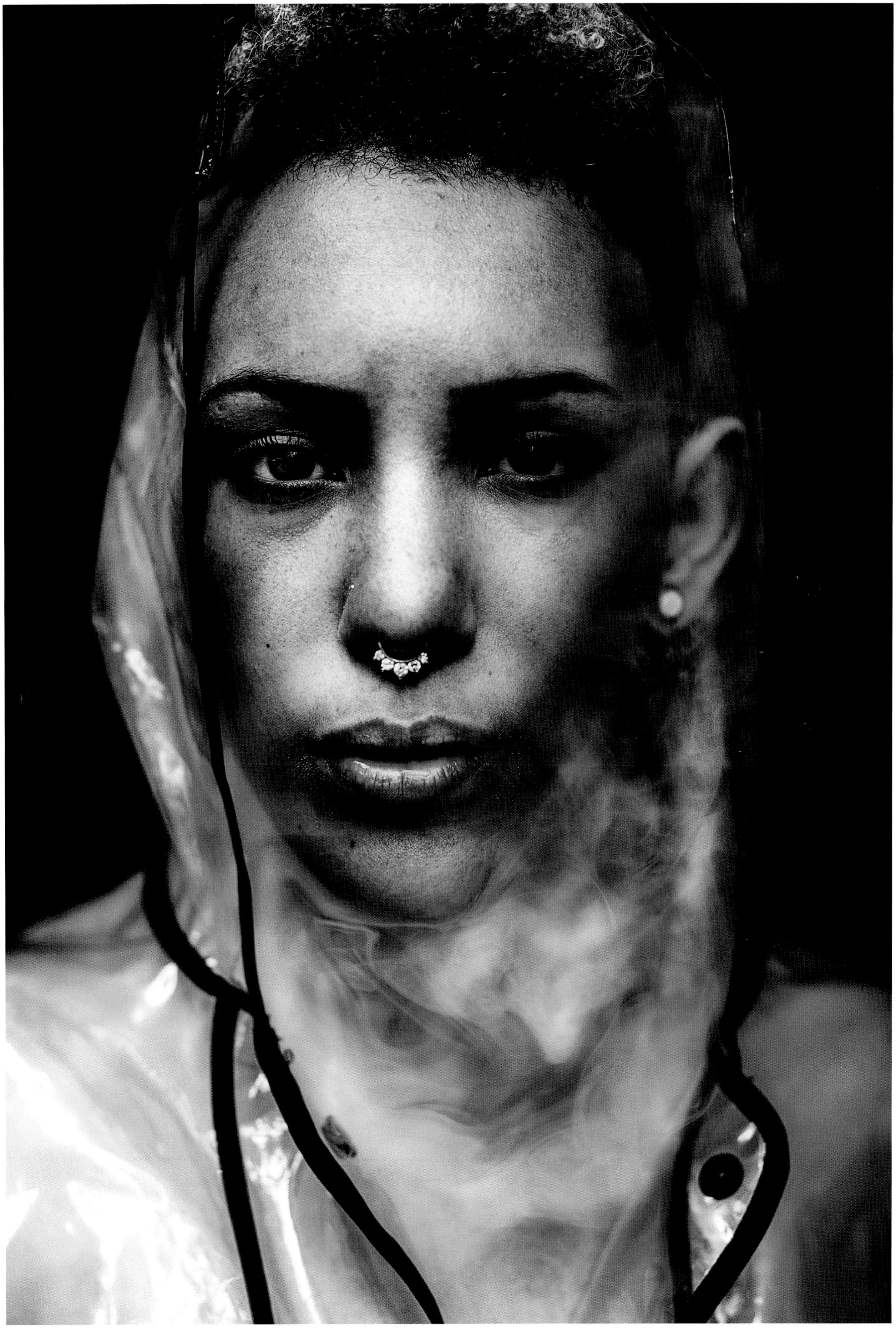

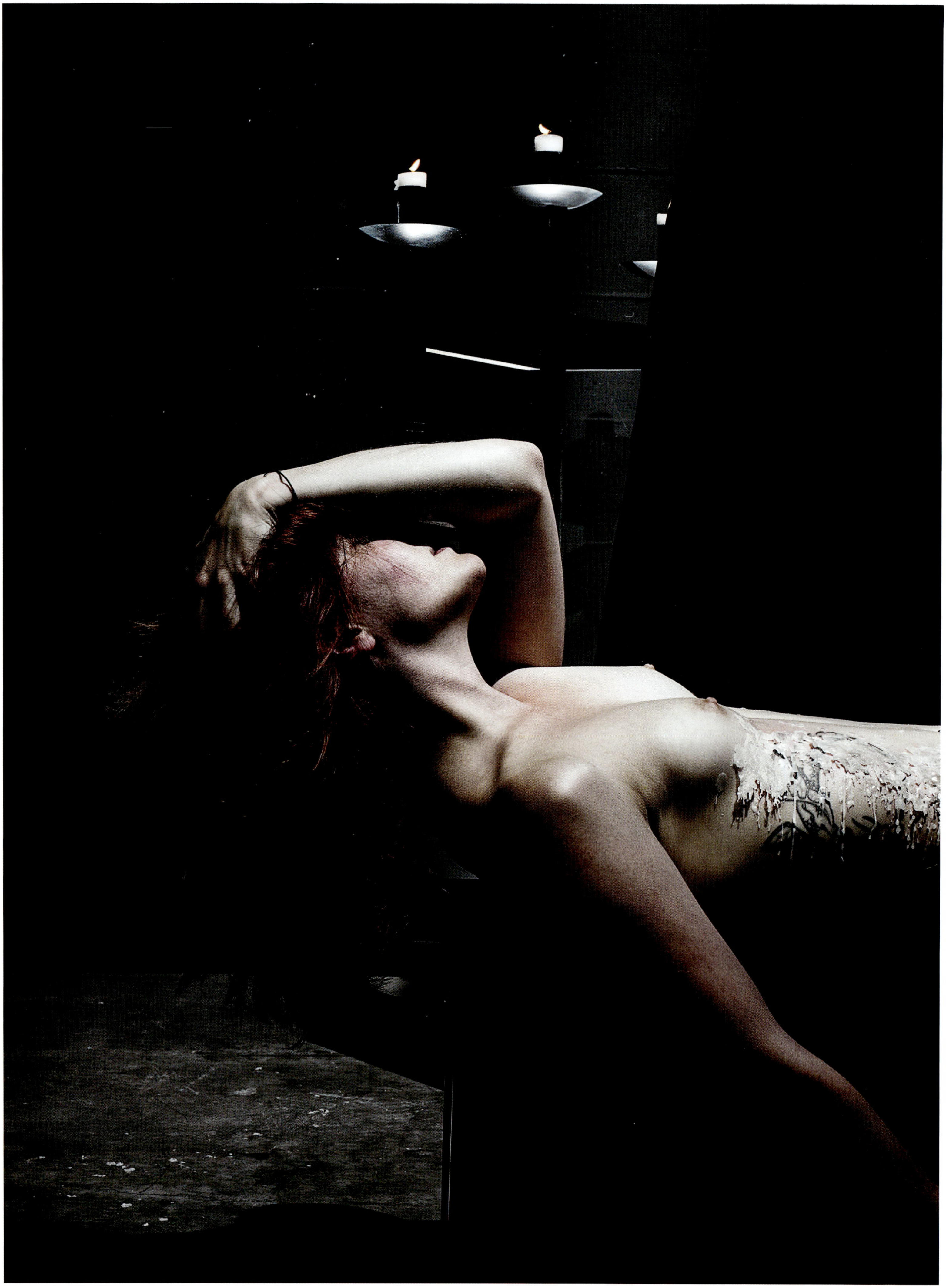

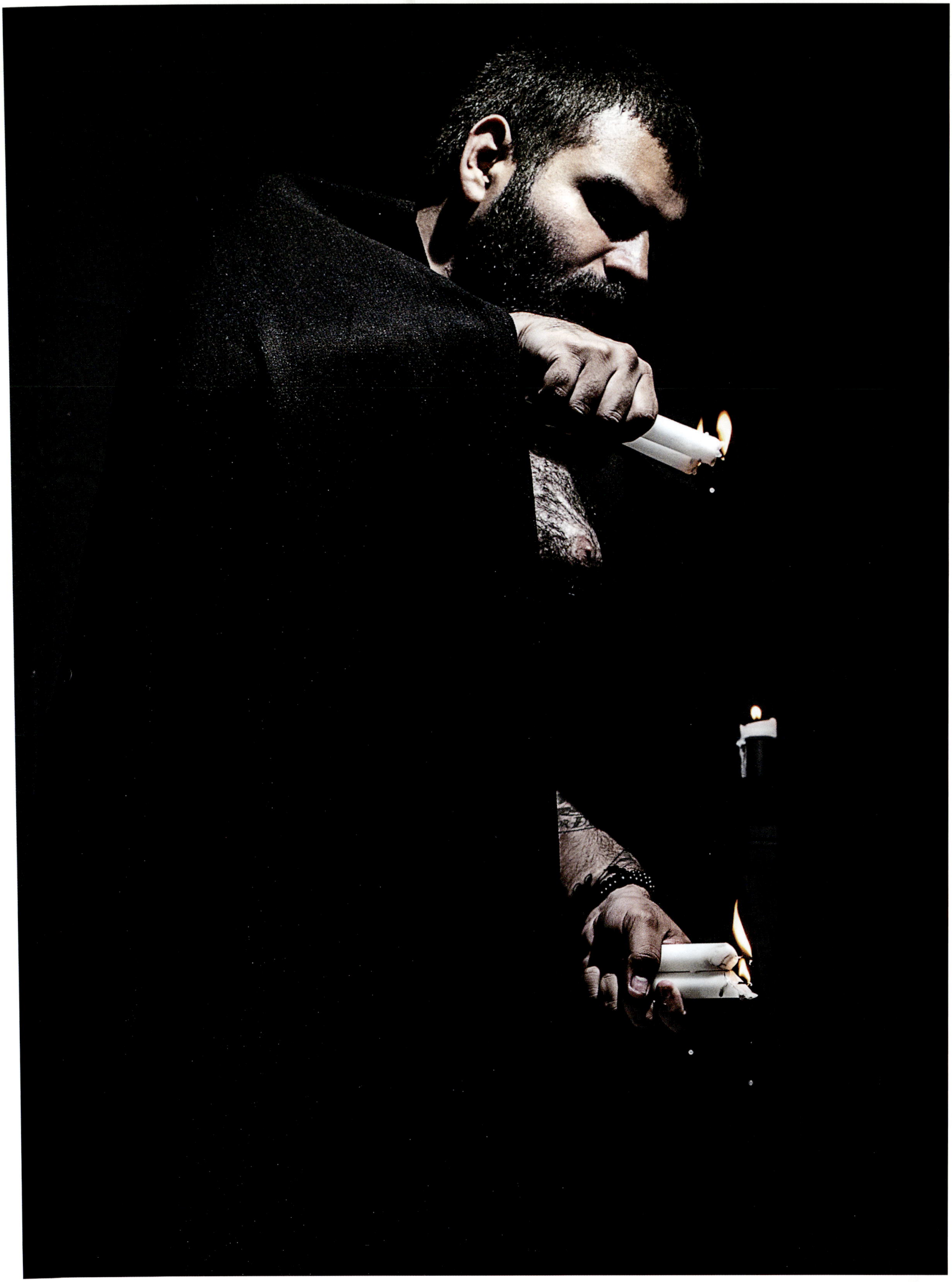

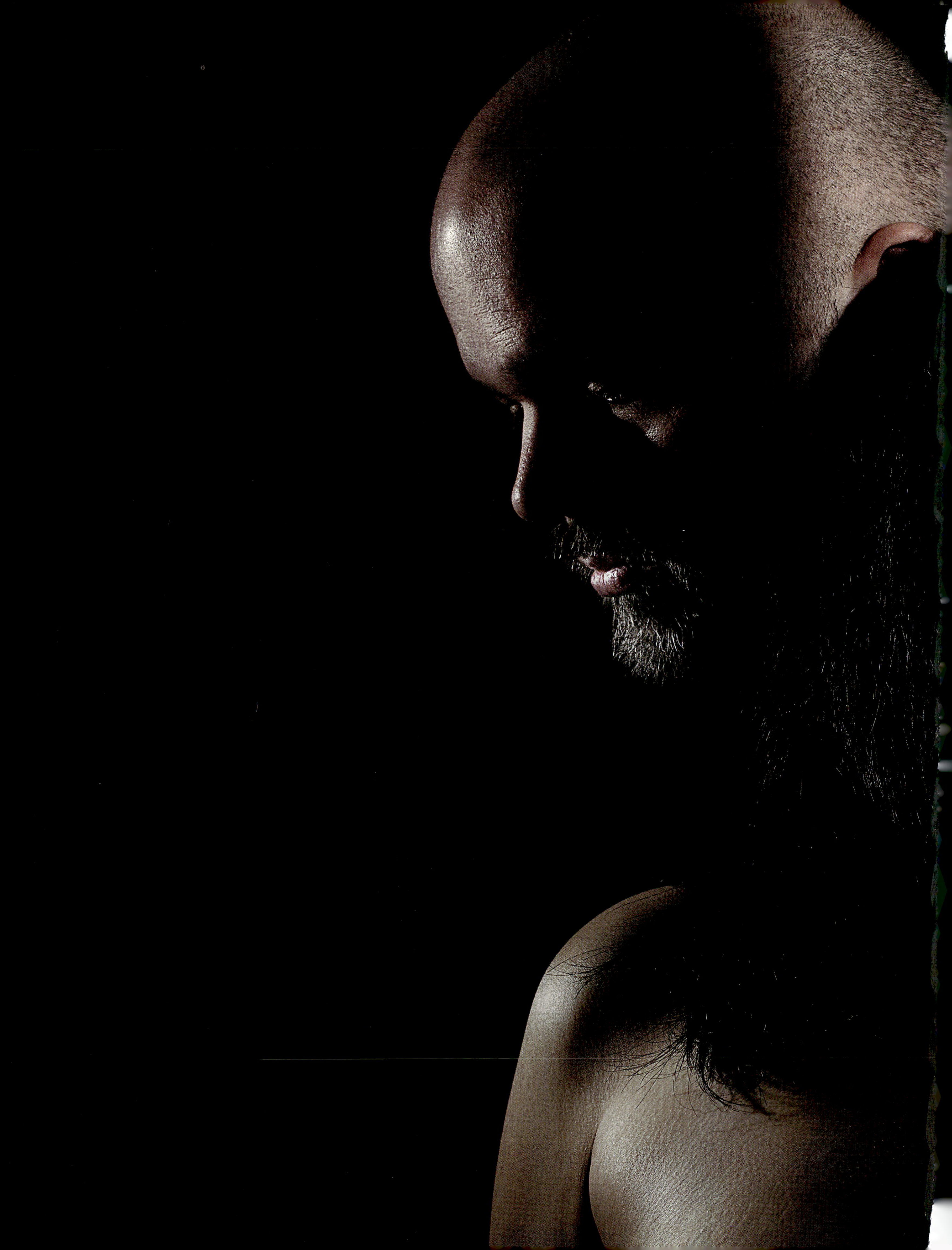

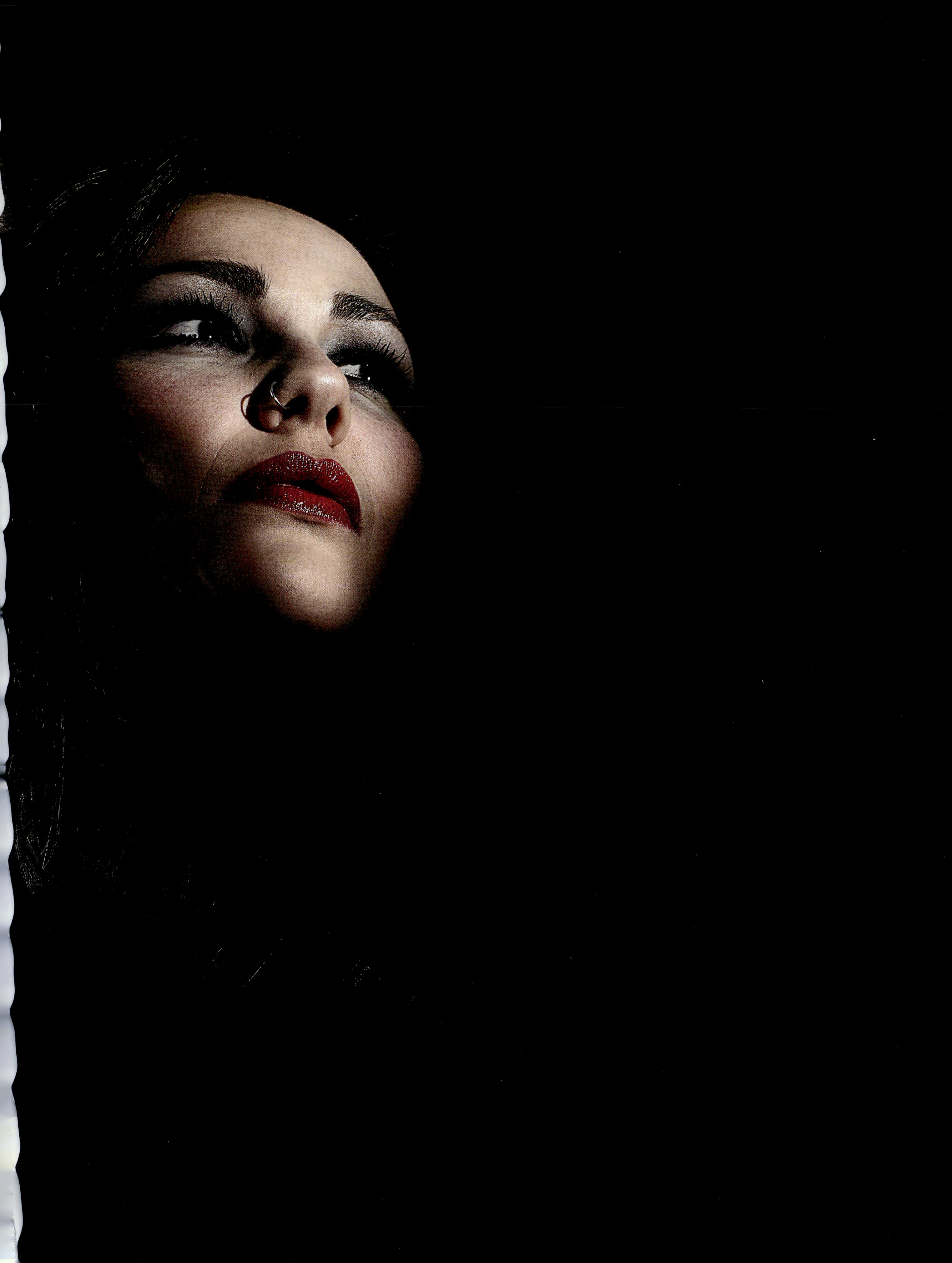

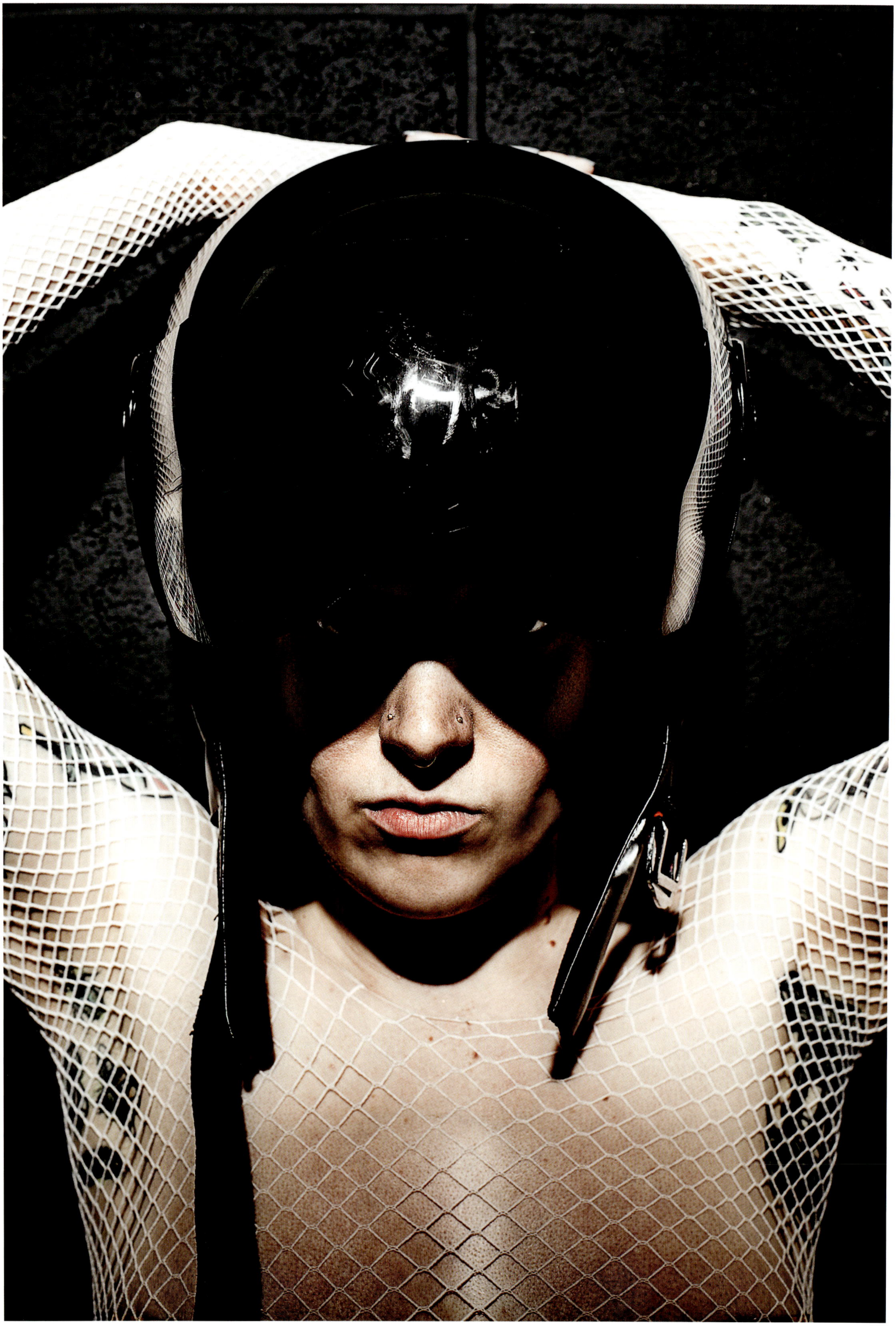

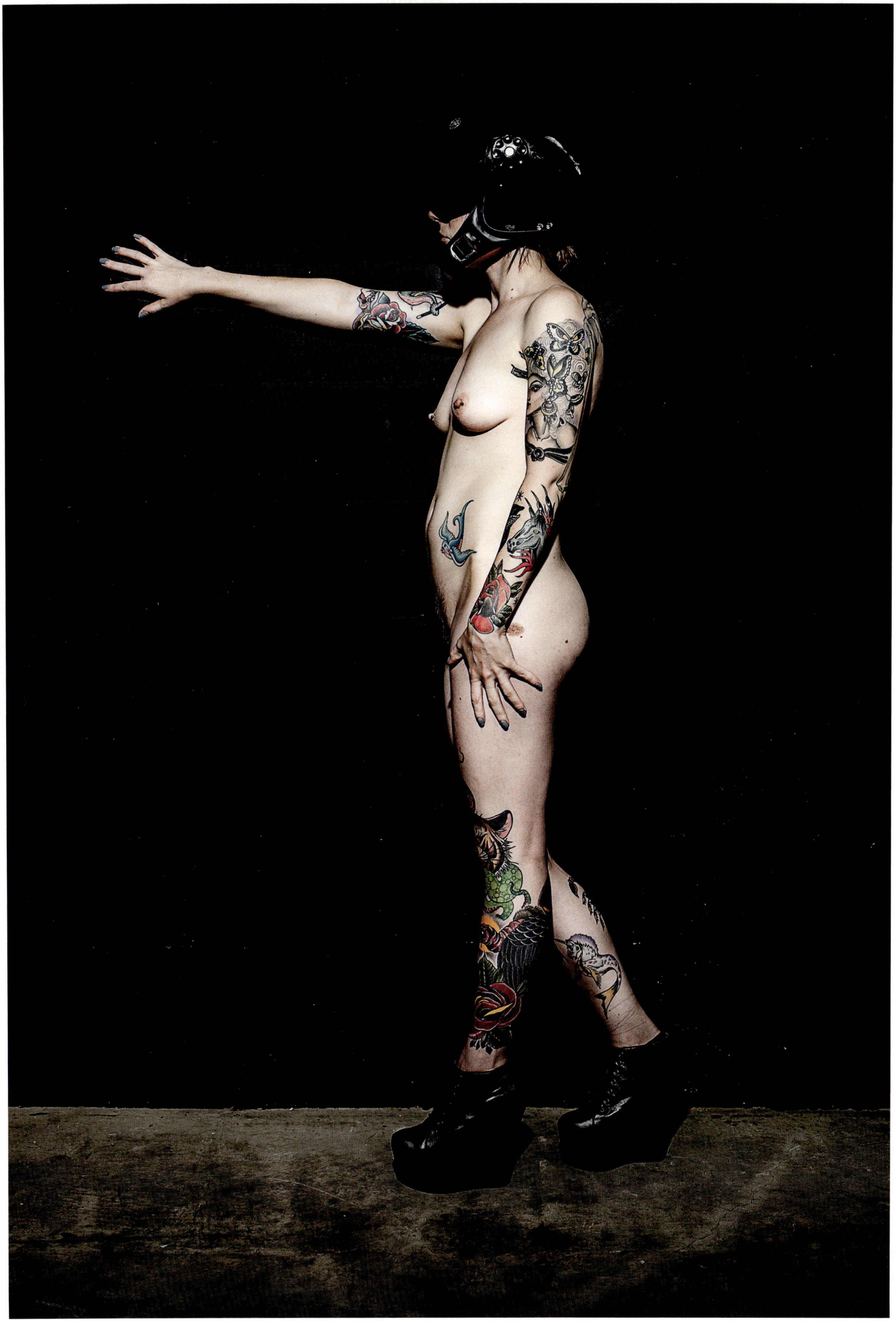

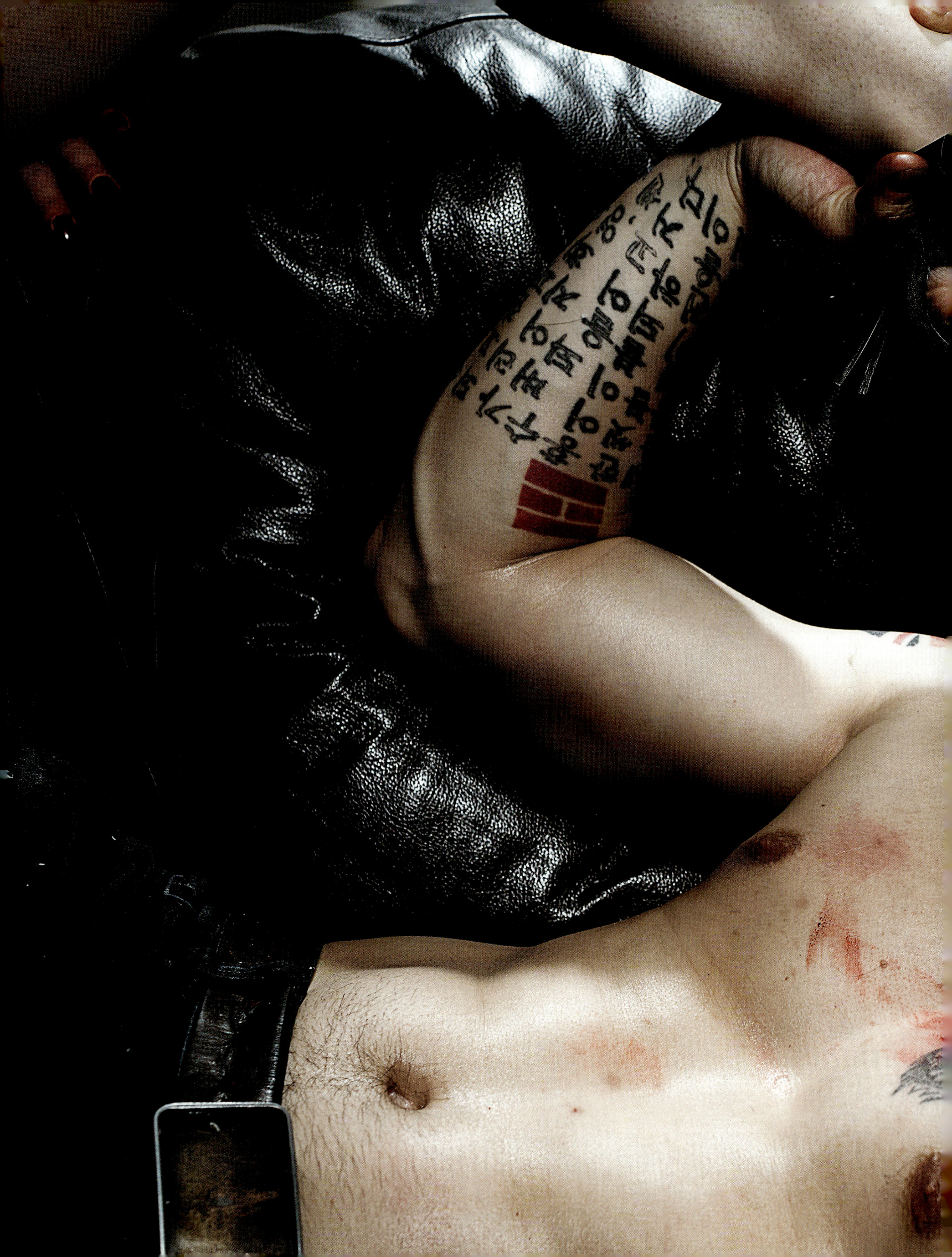

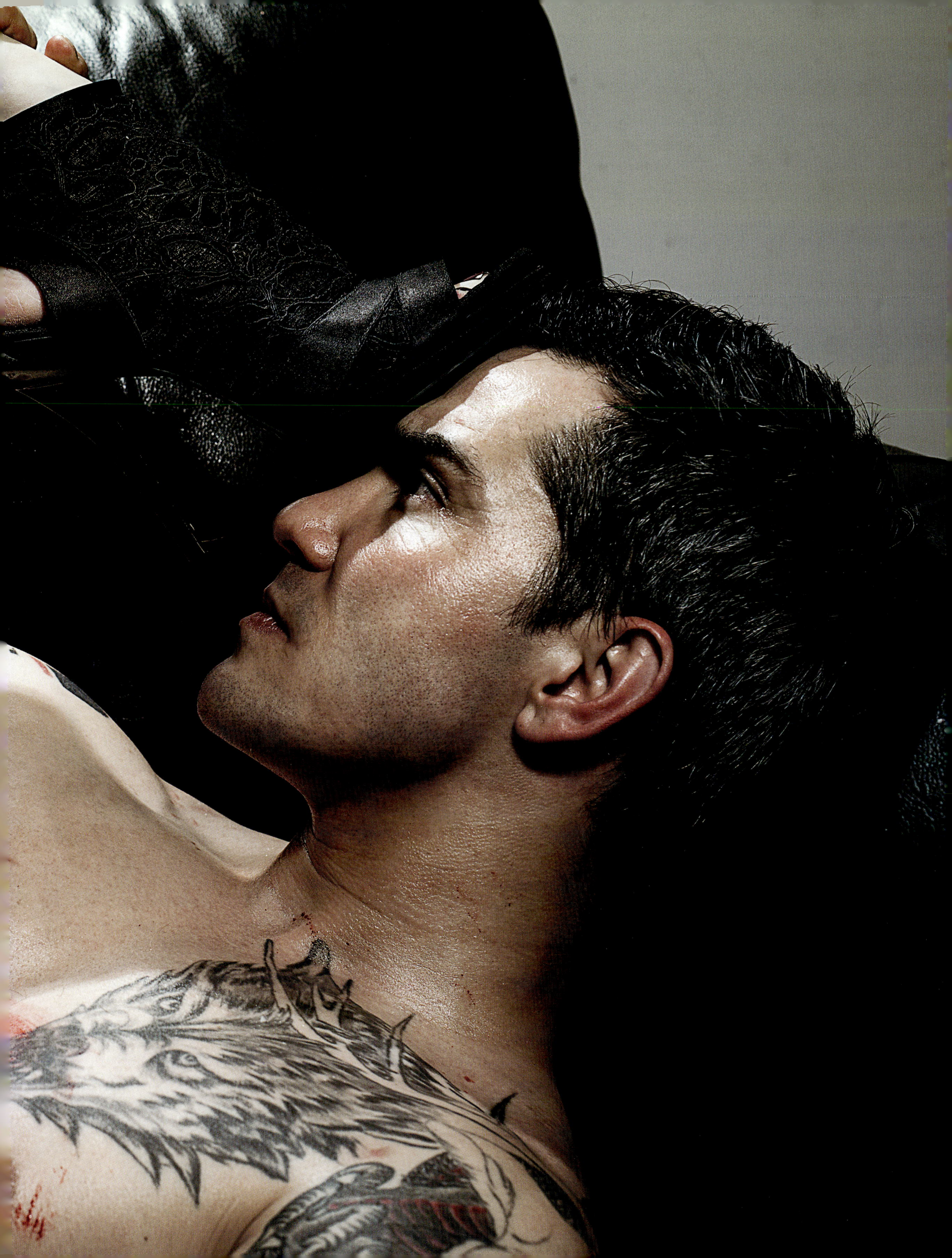

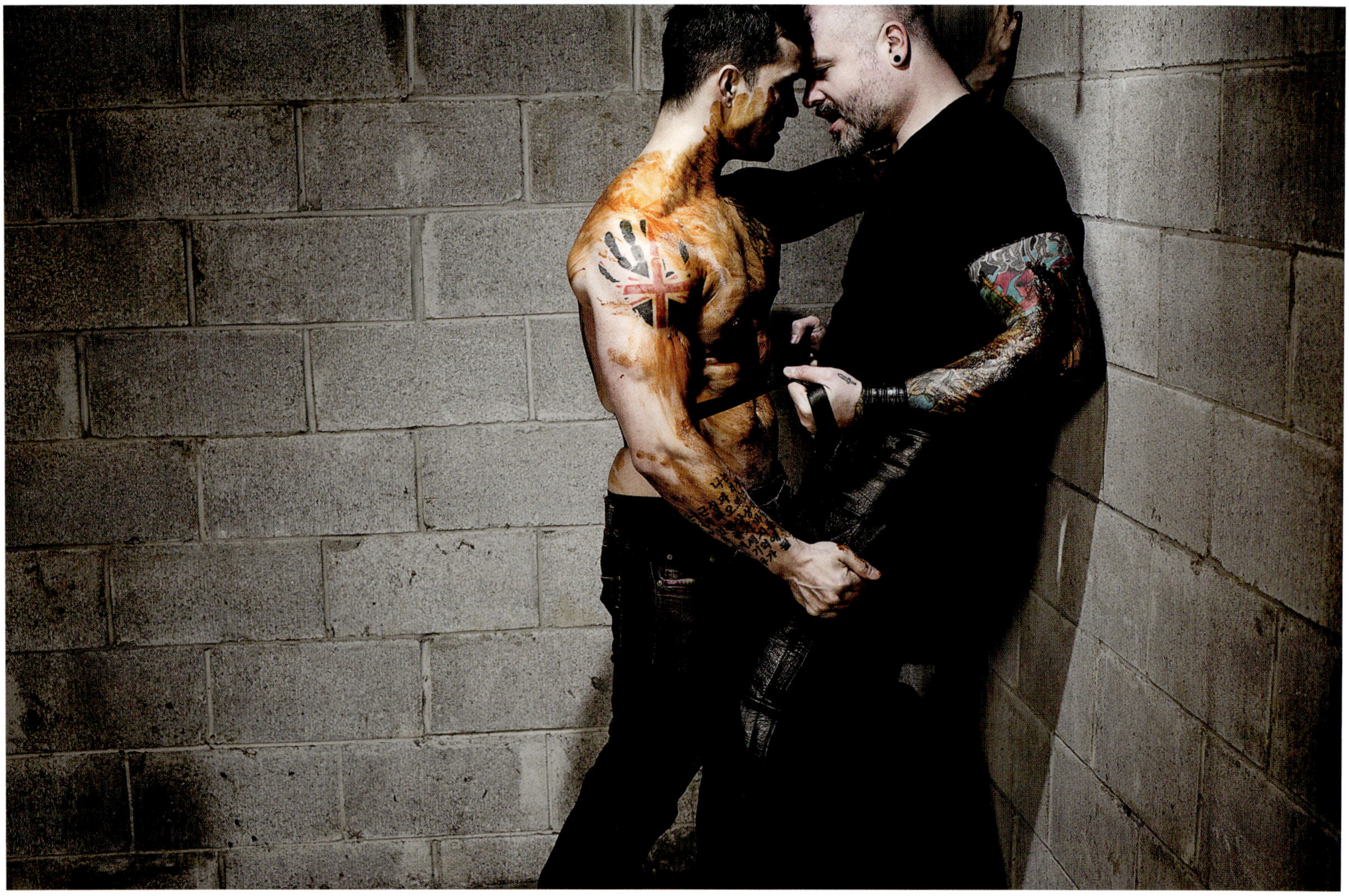

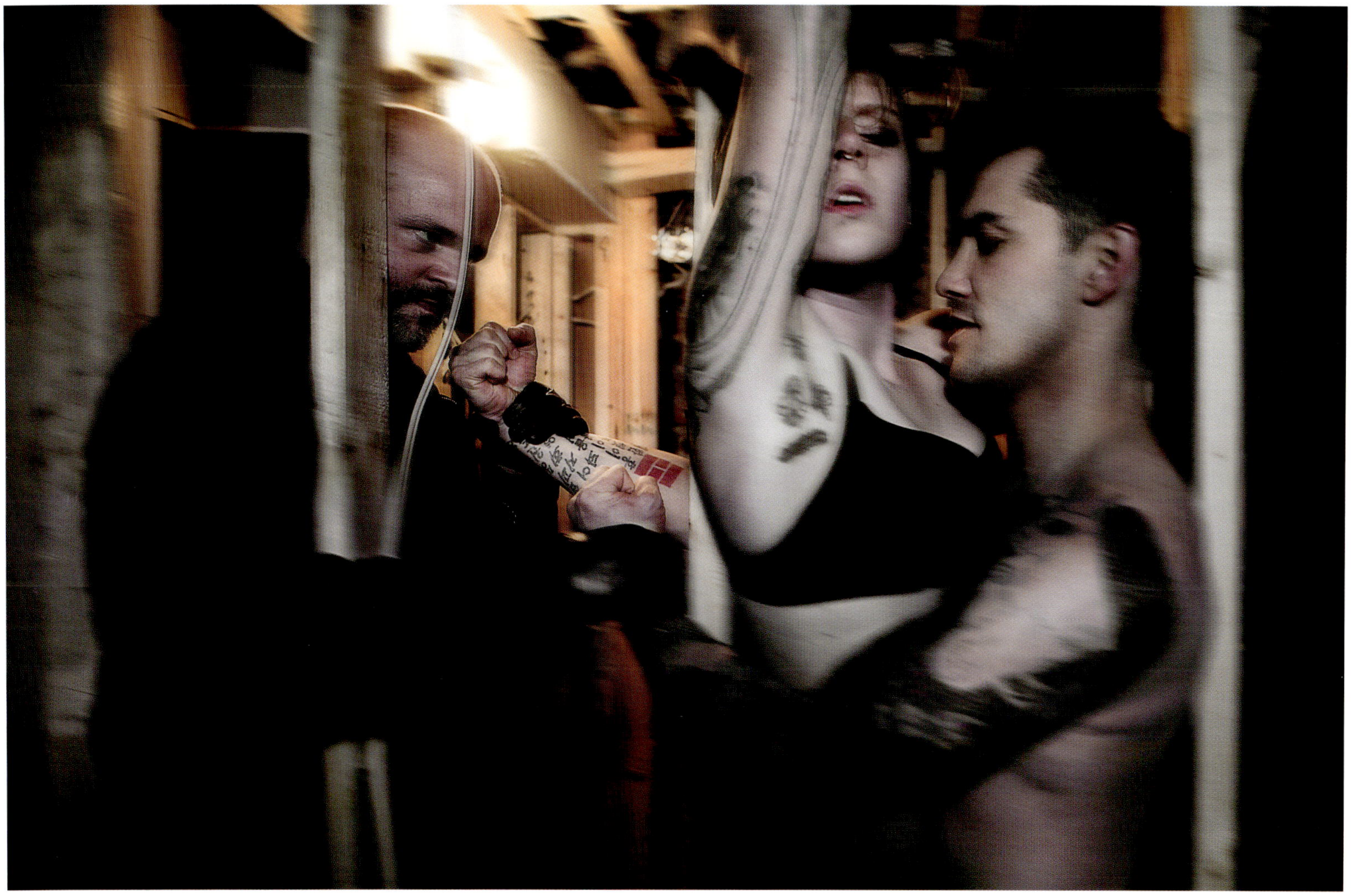

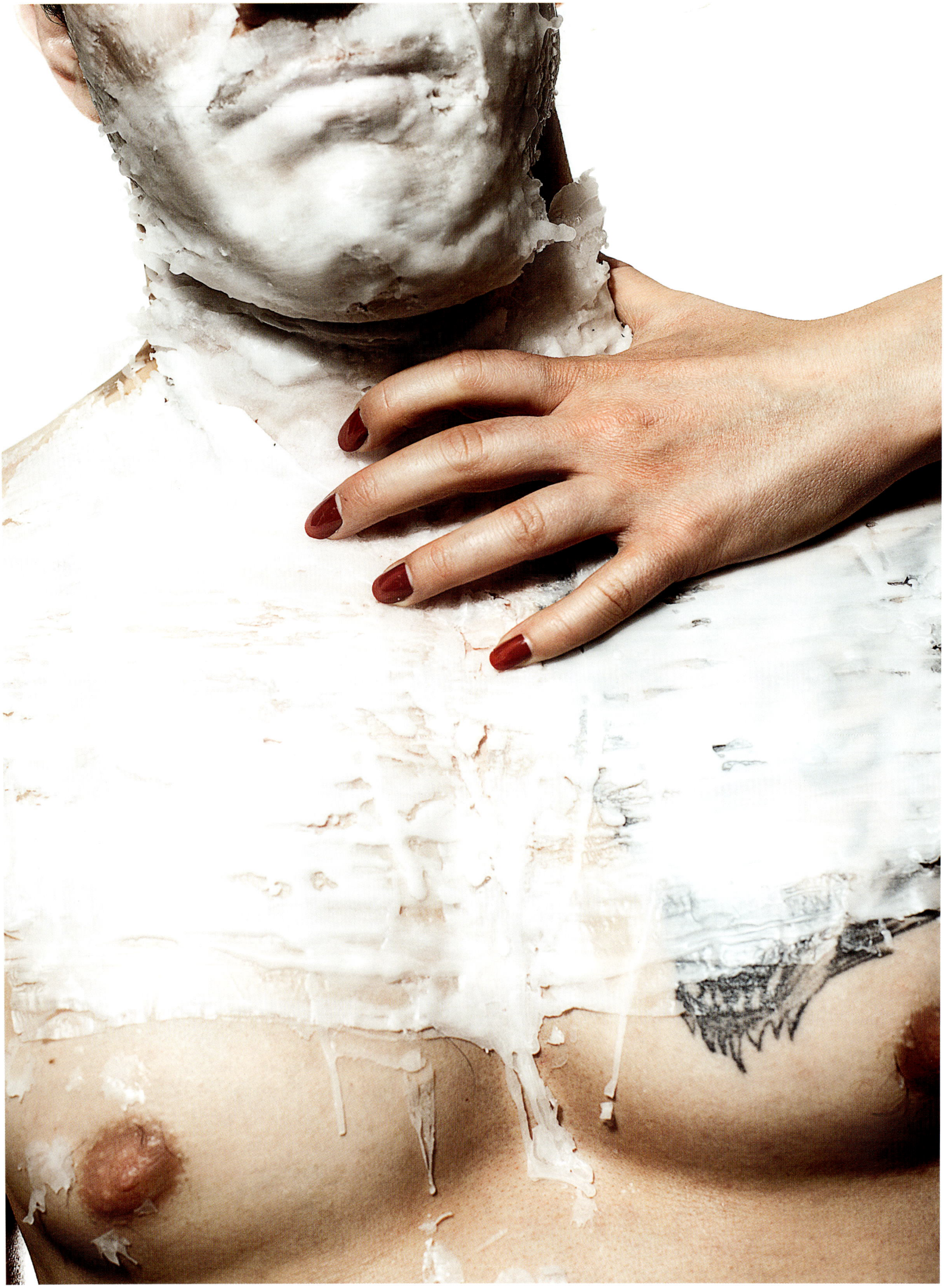

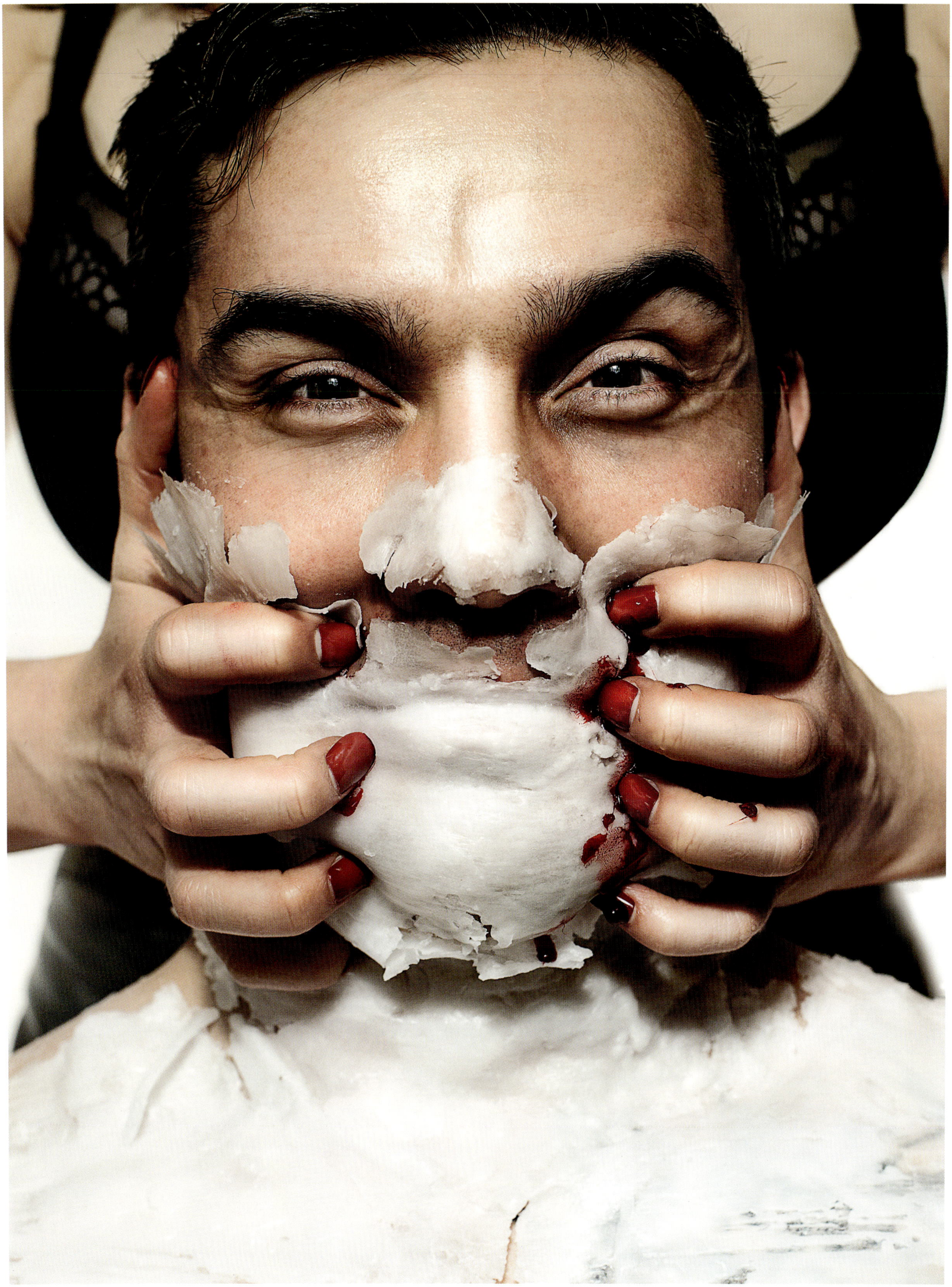

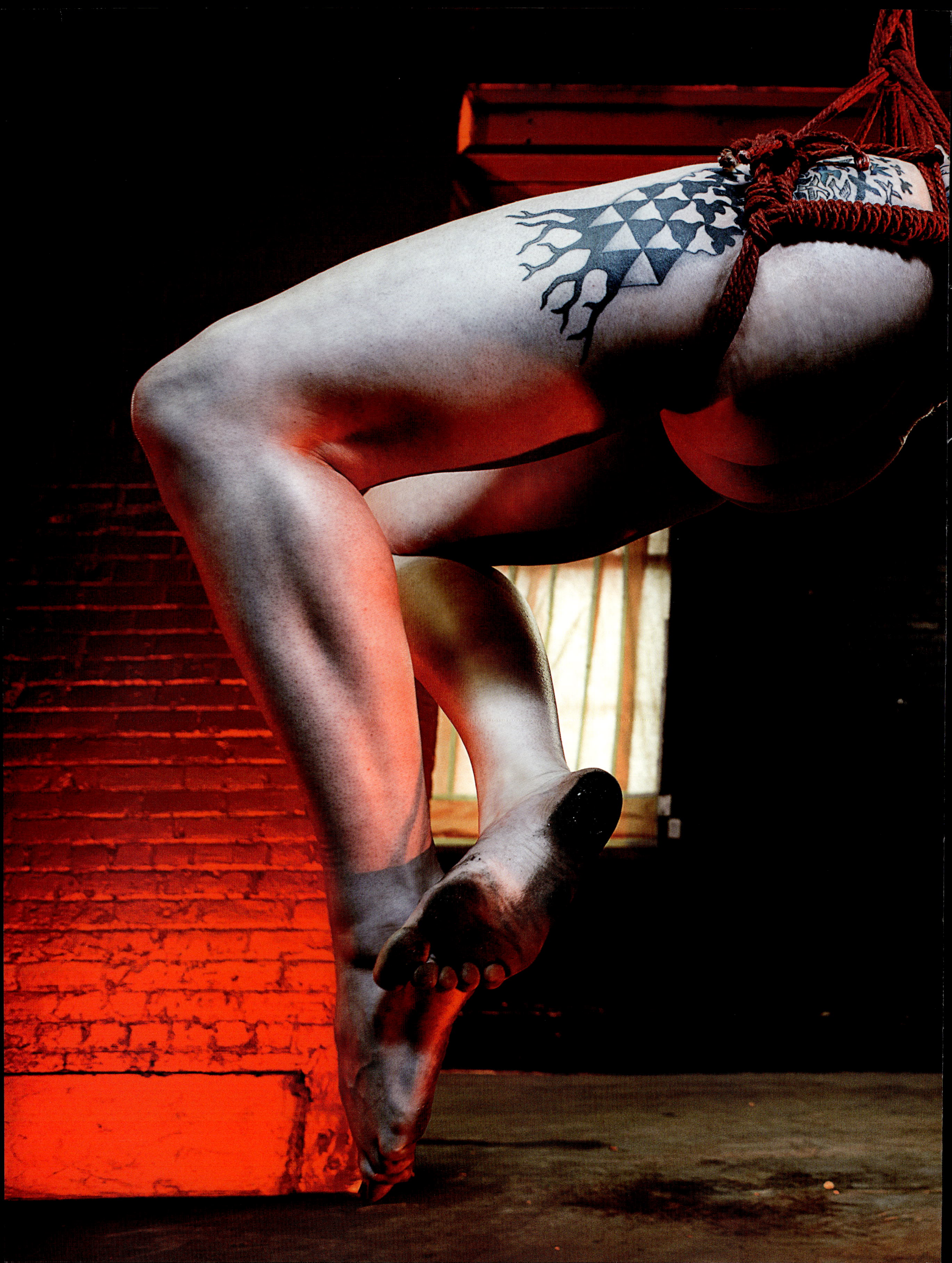

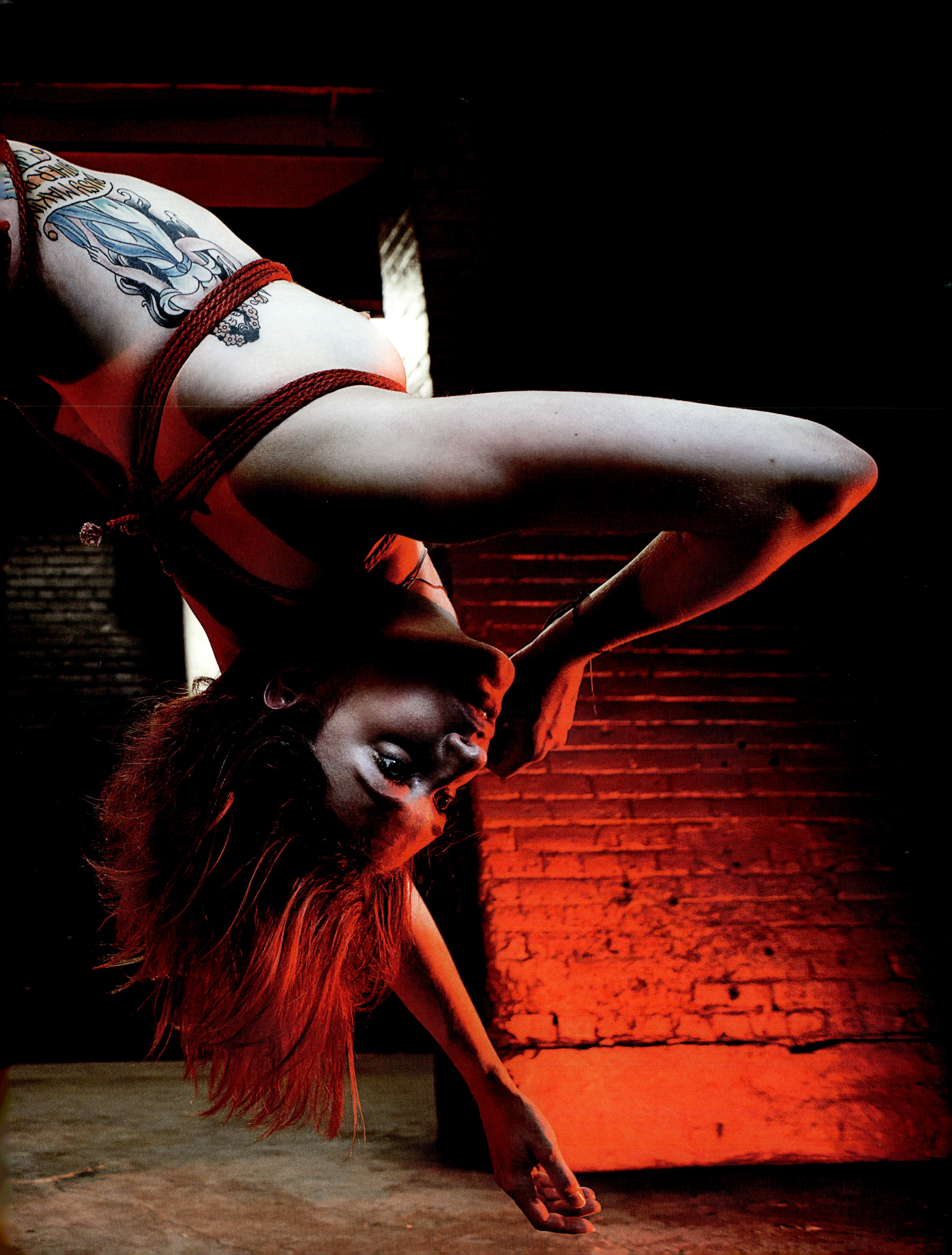

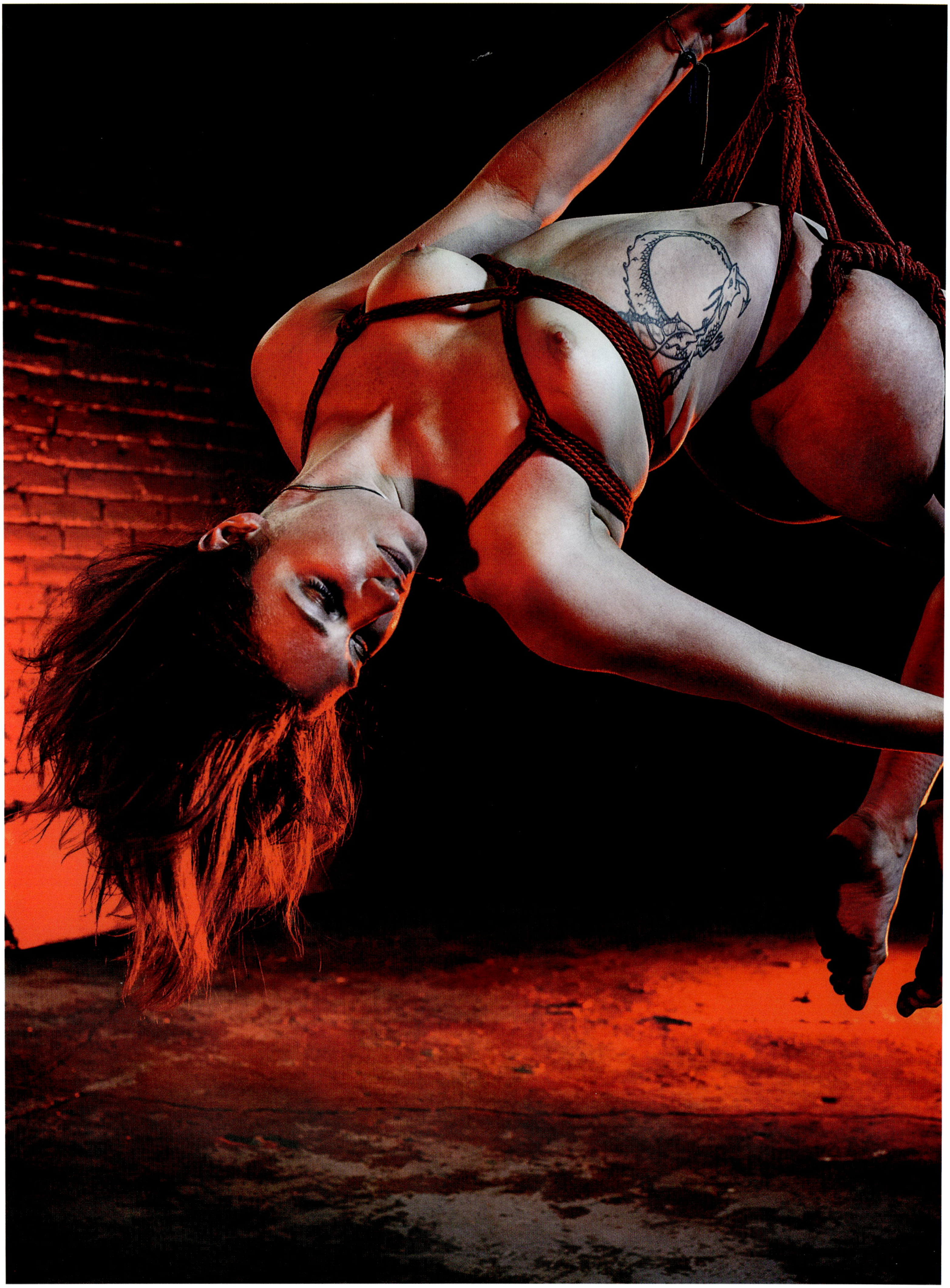

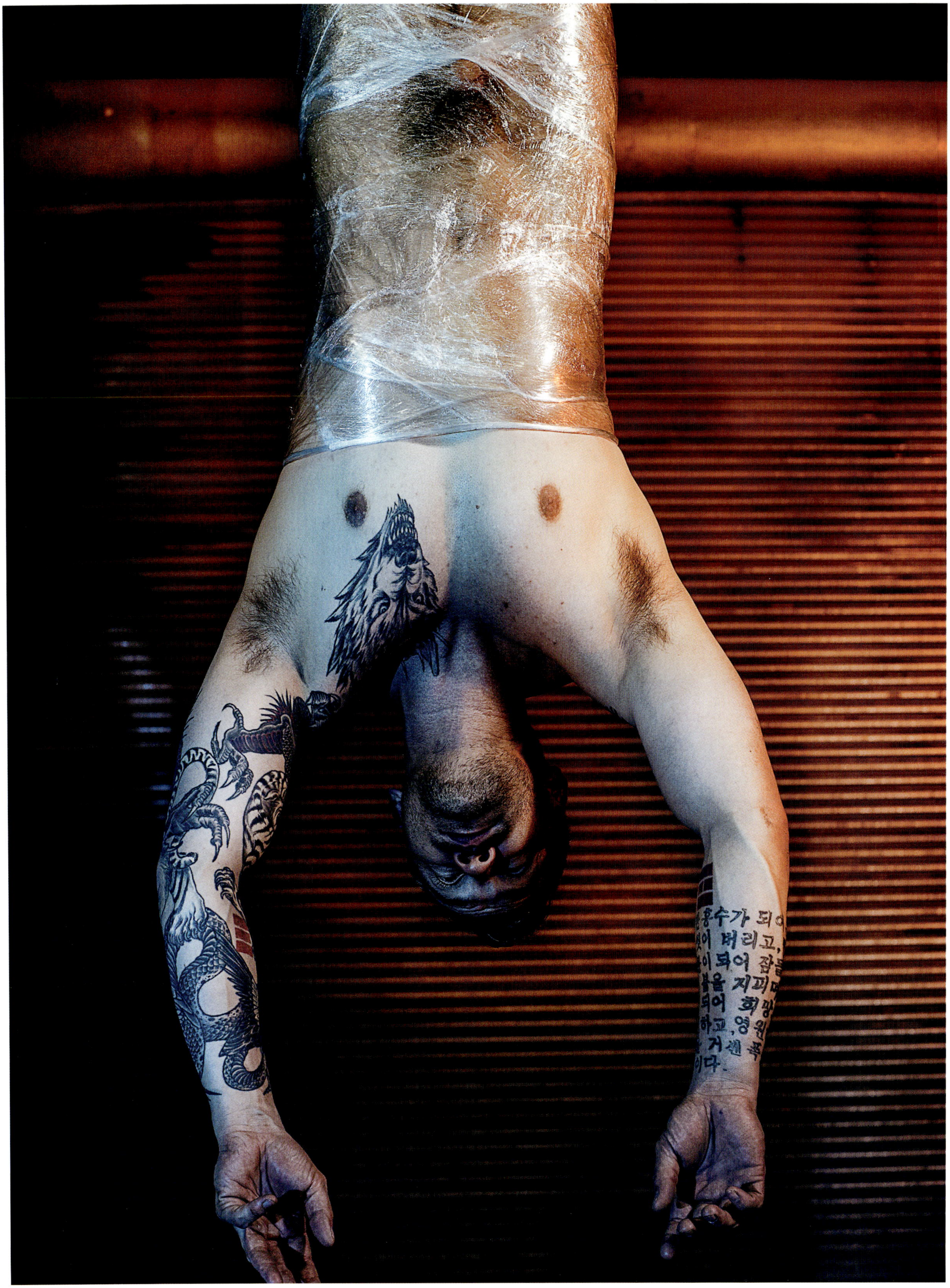

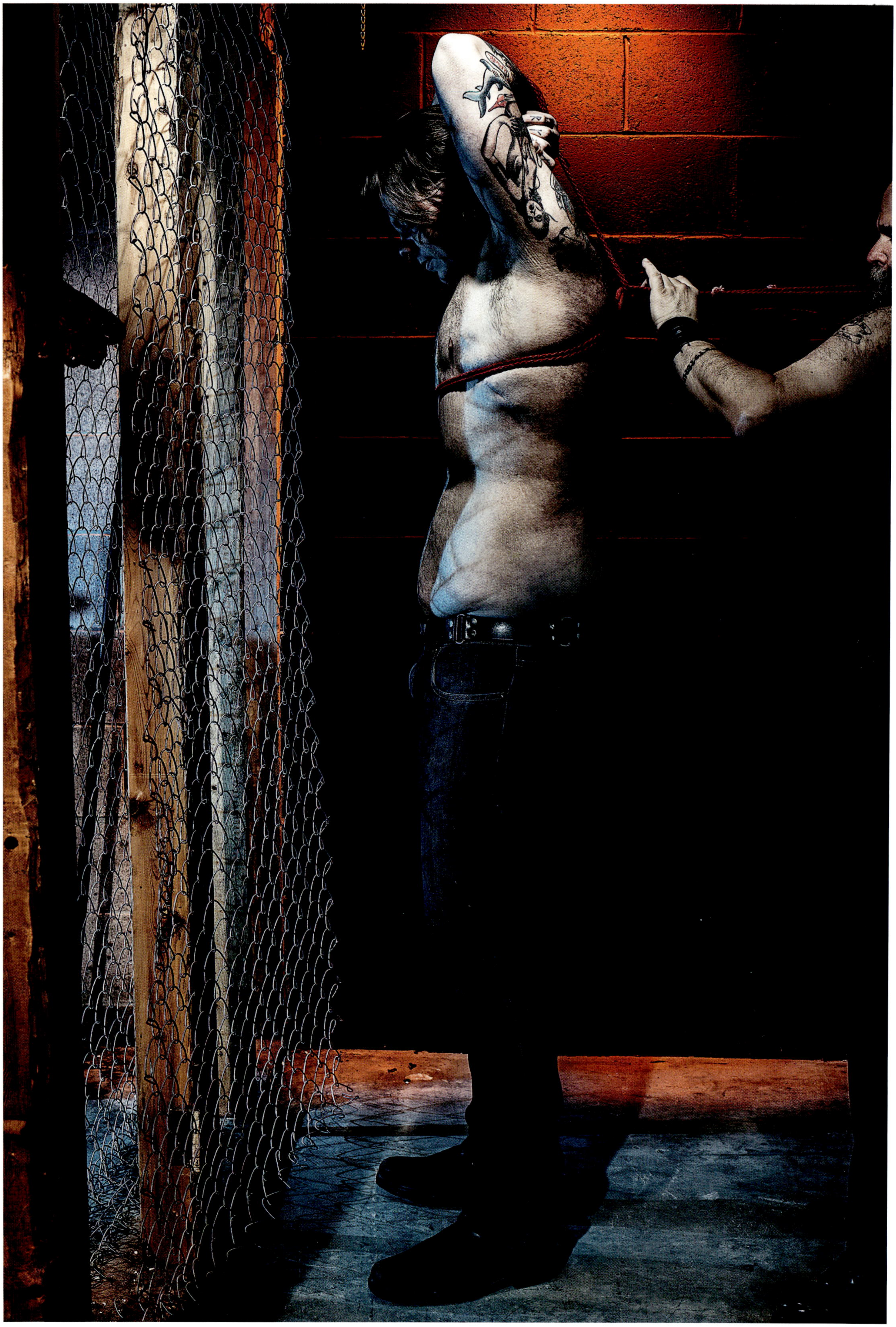

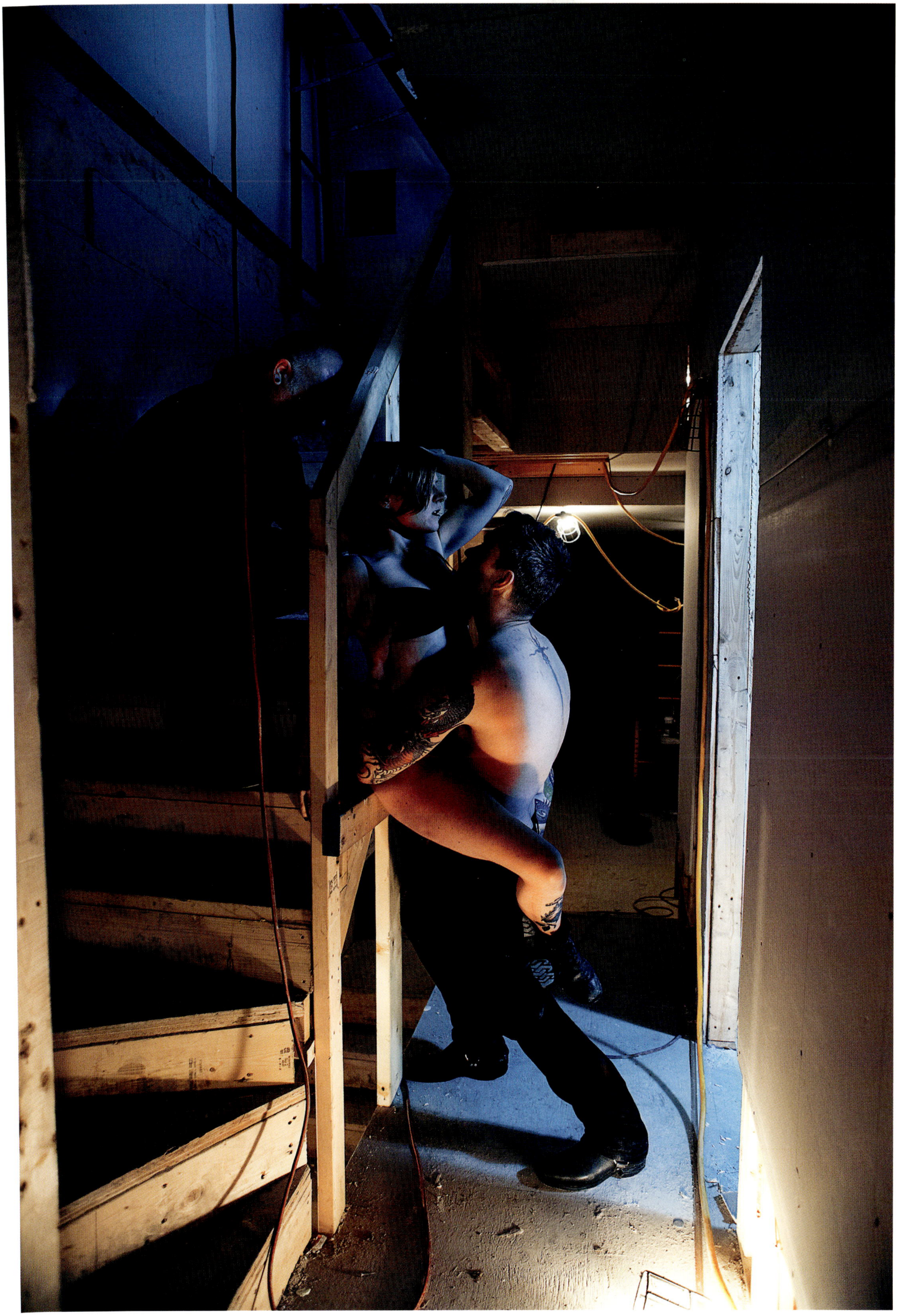

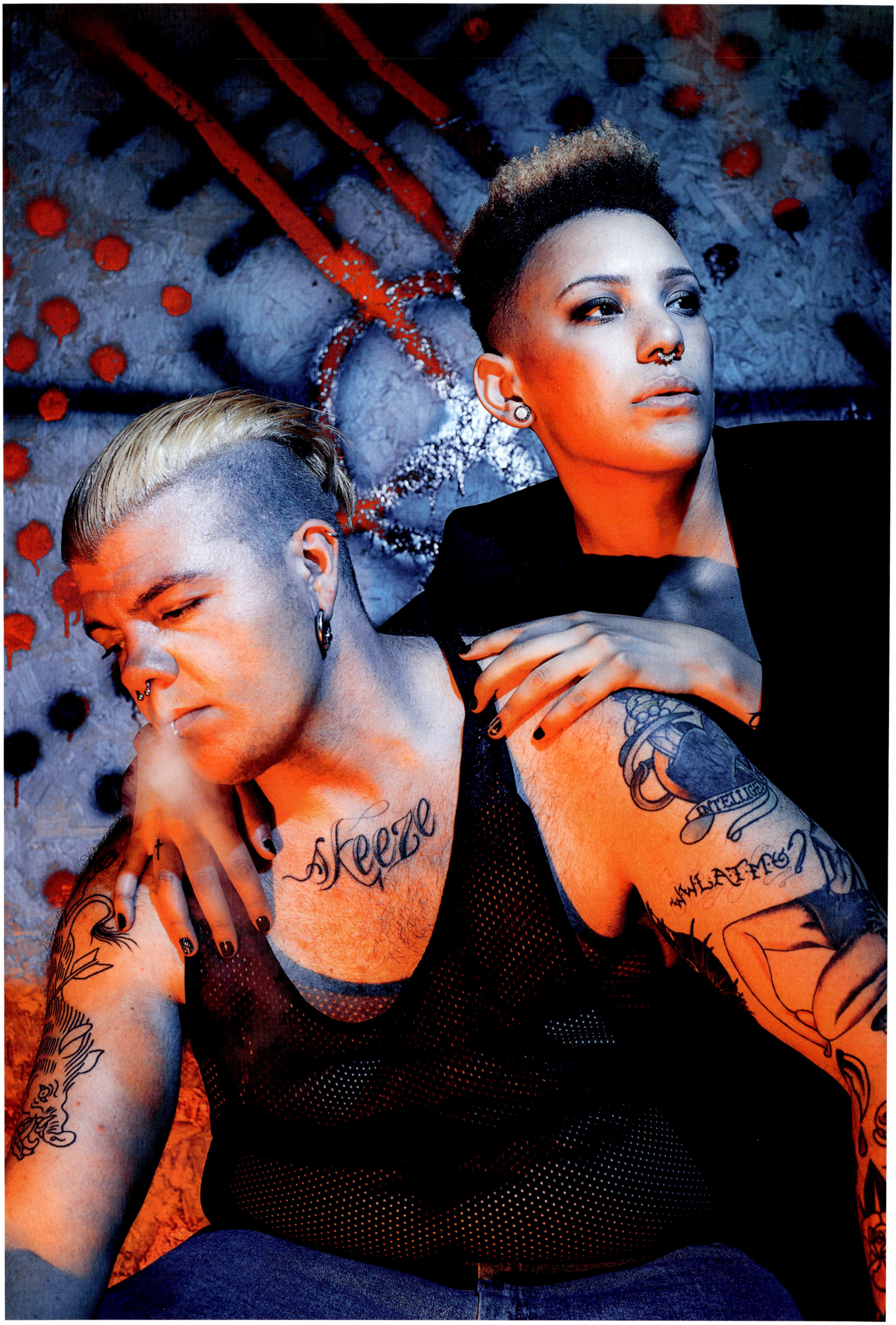
skeeze

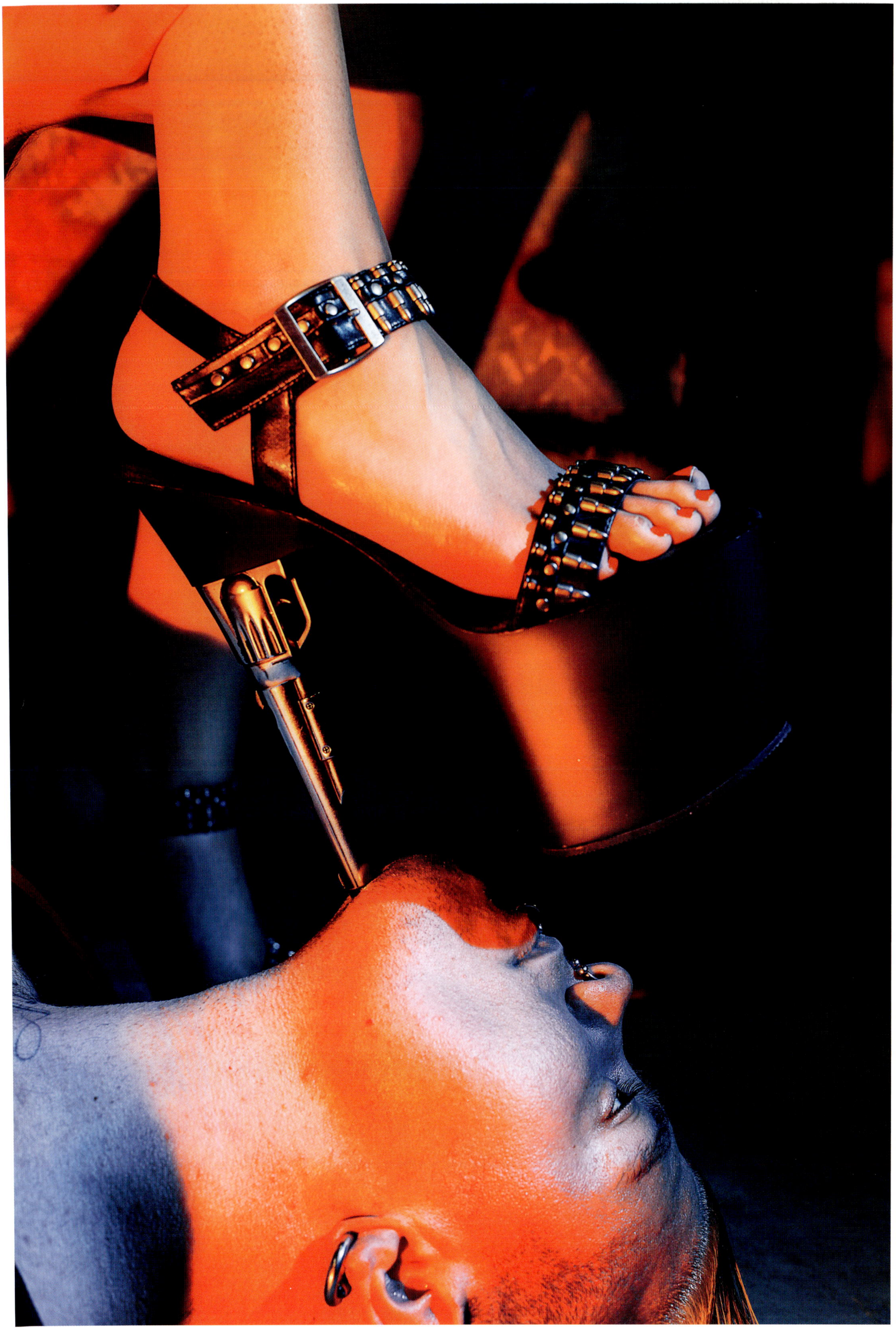

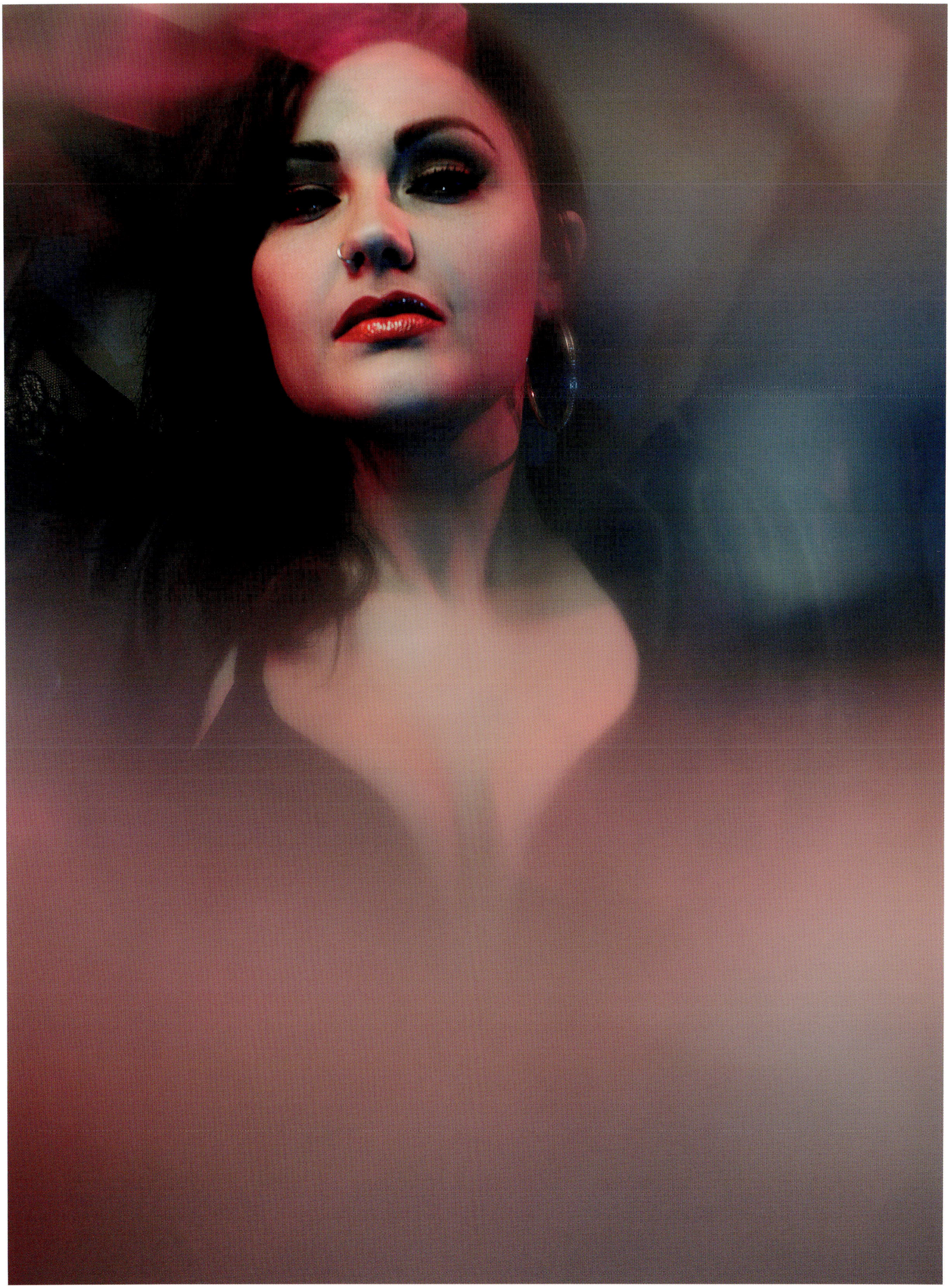

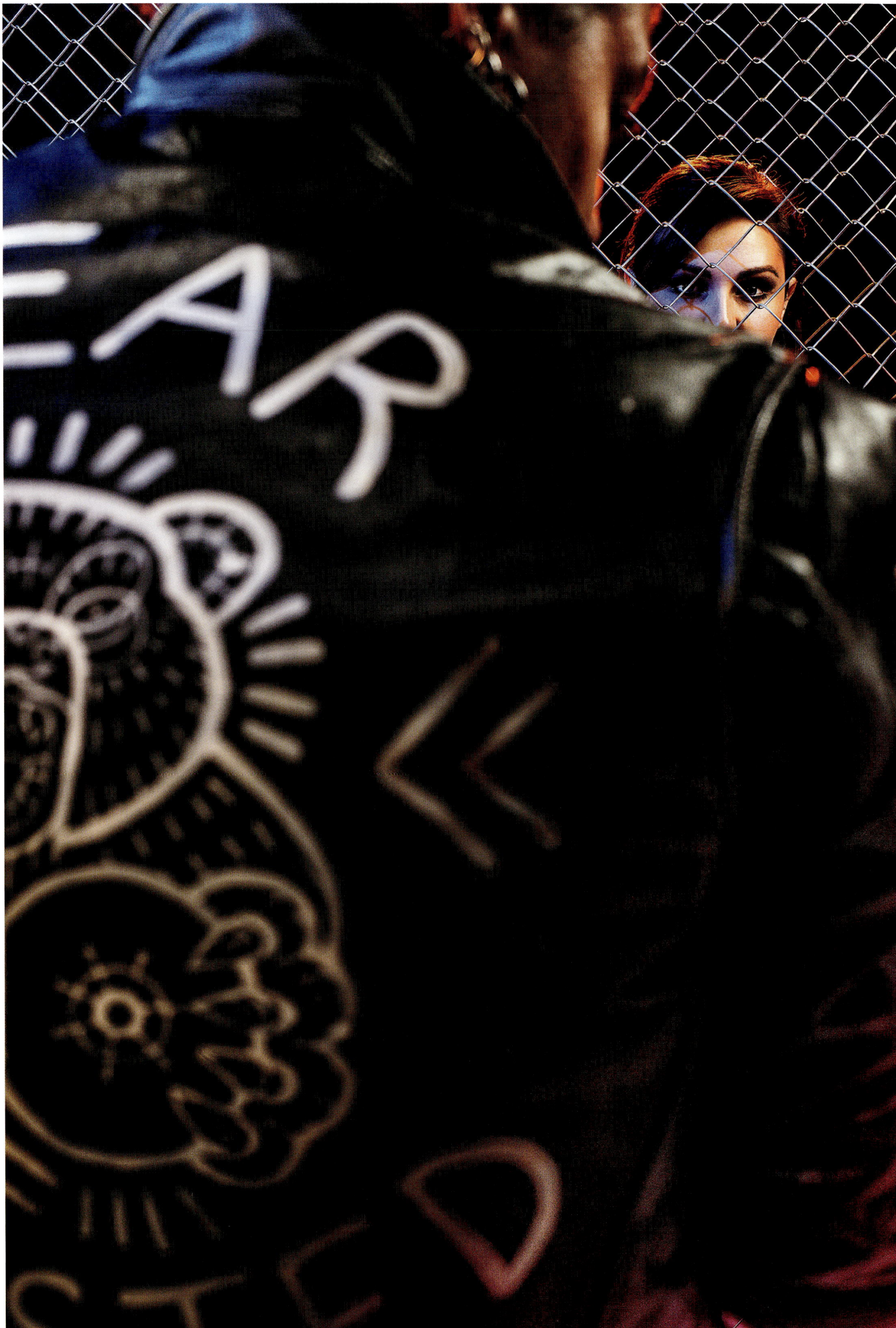

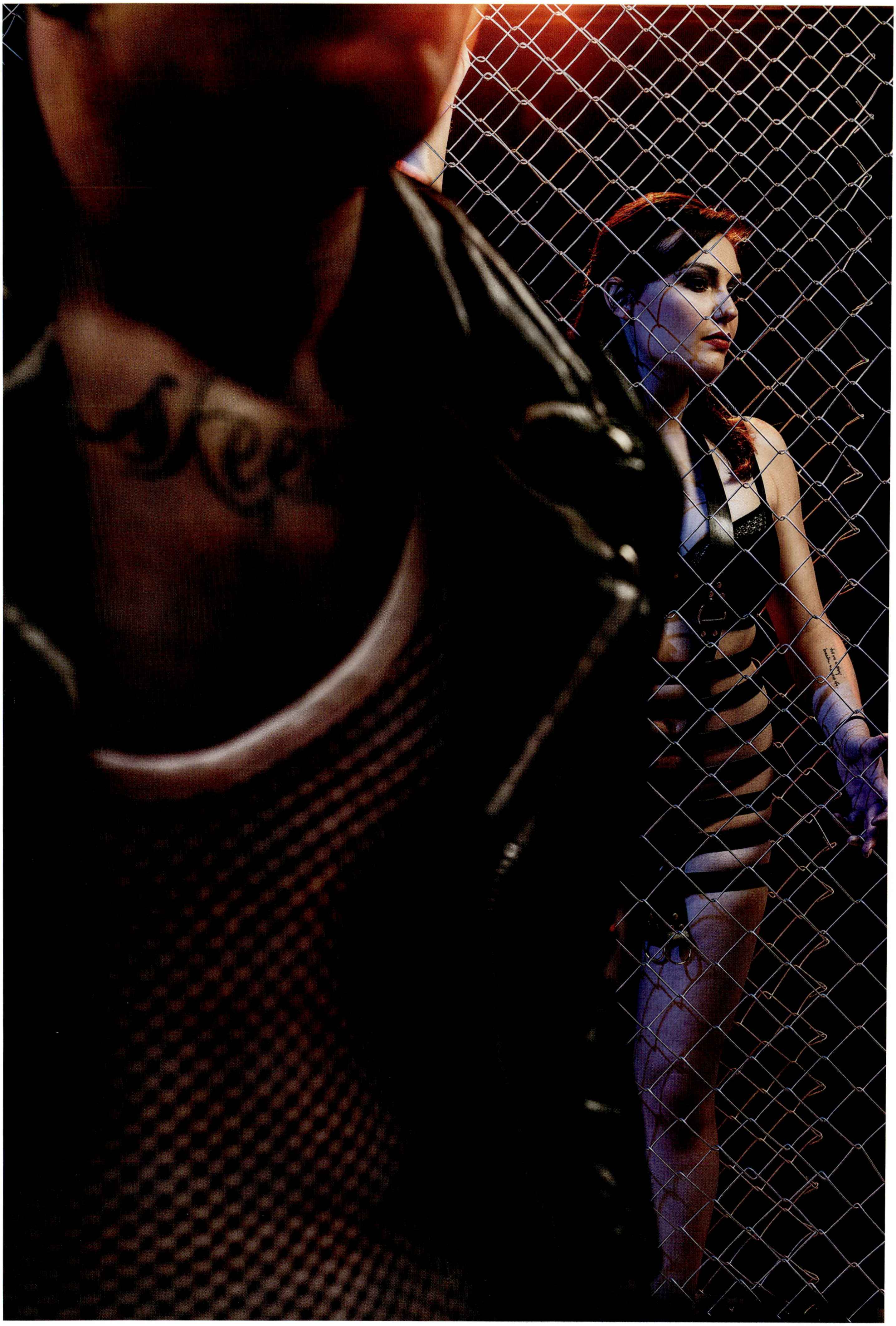

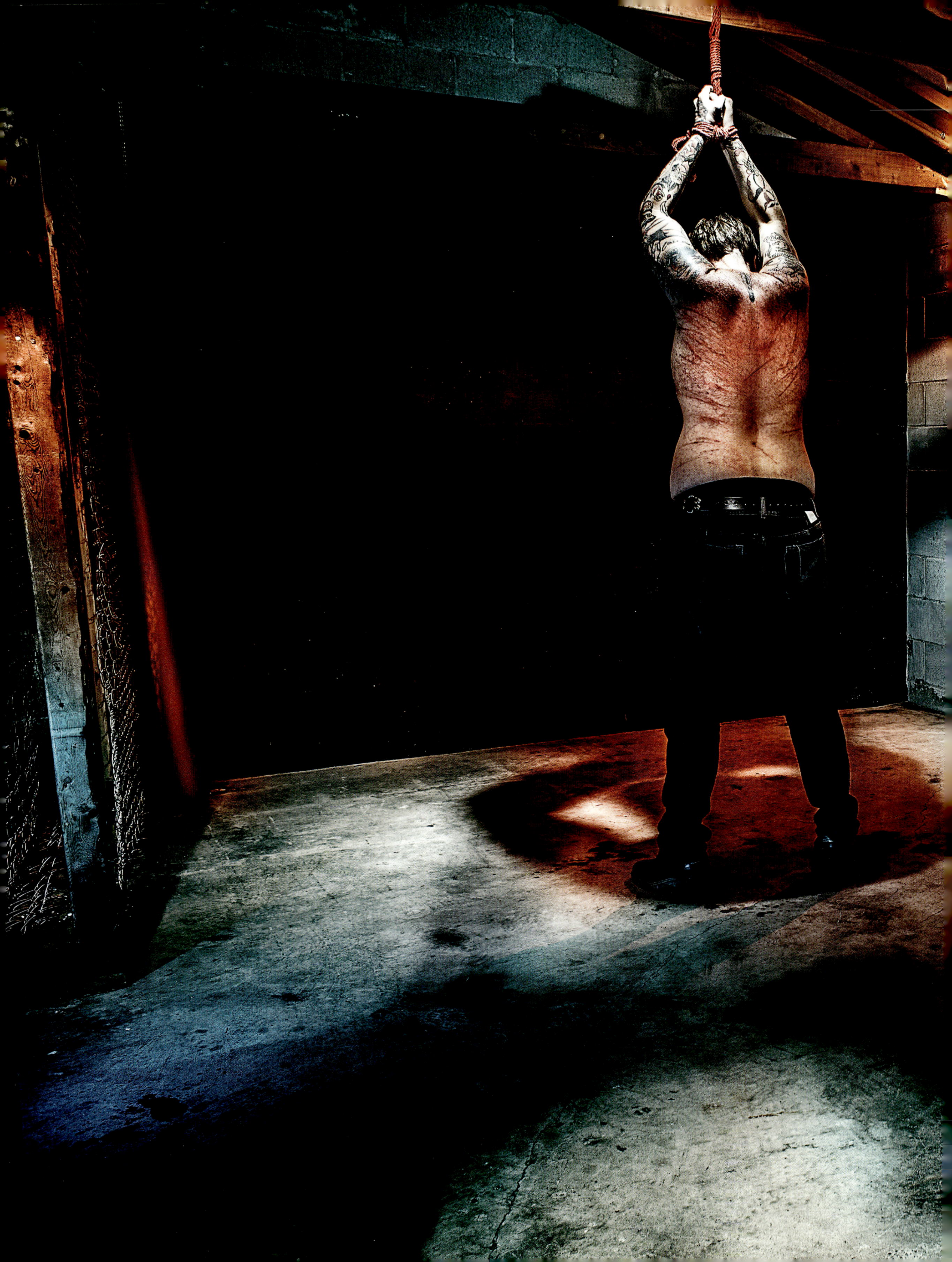

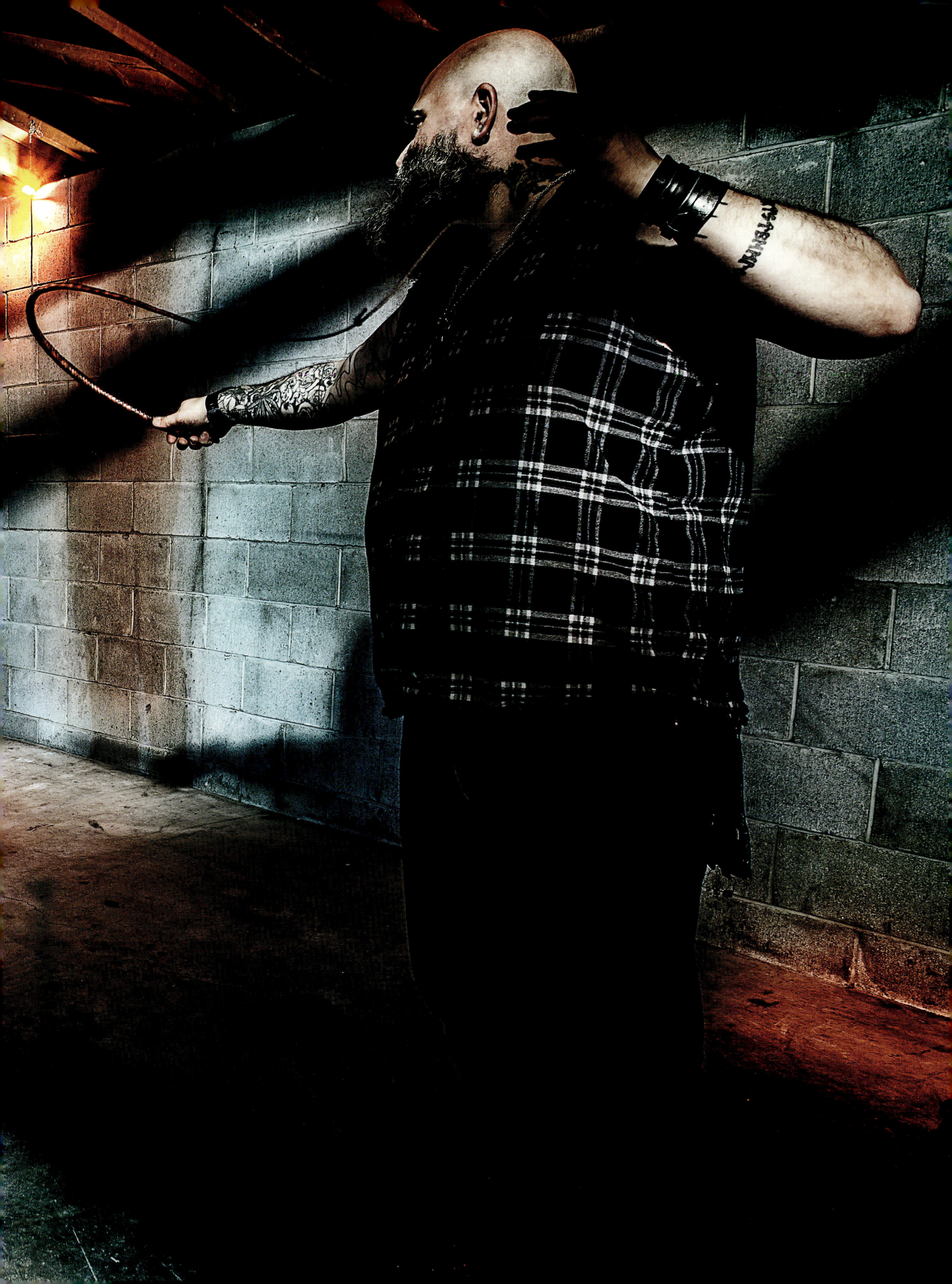

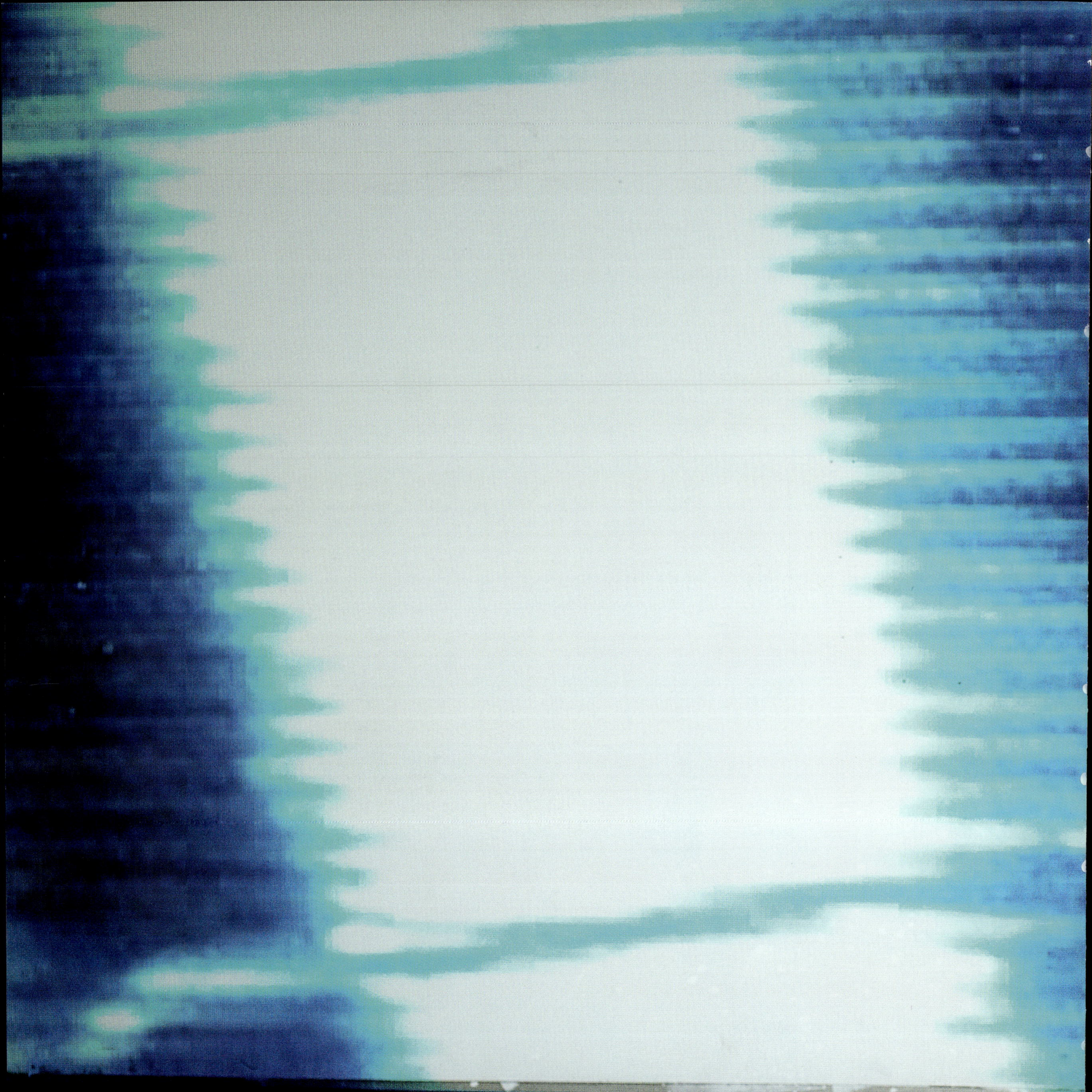

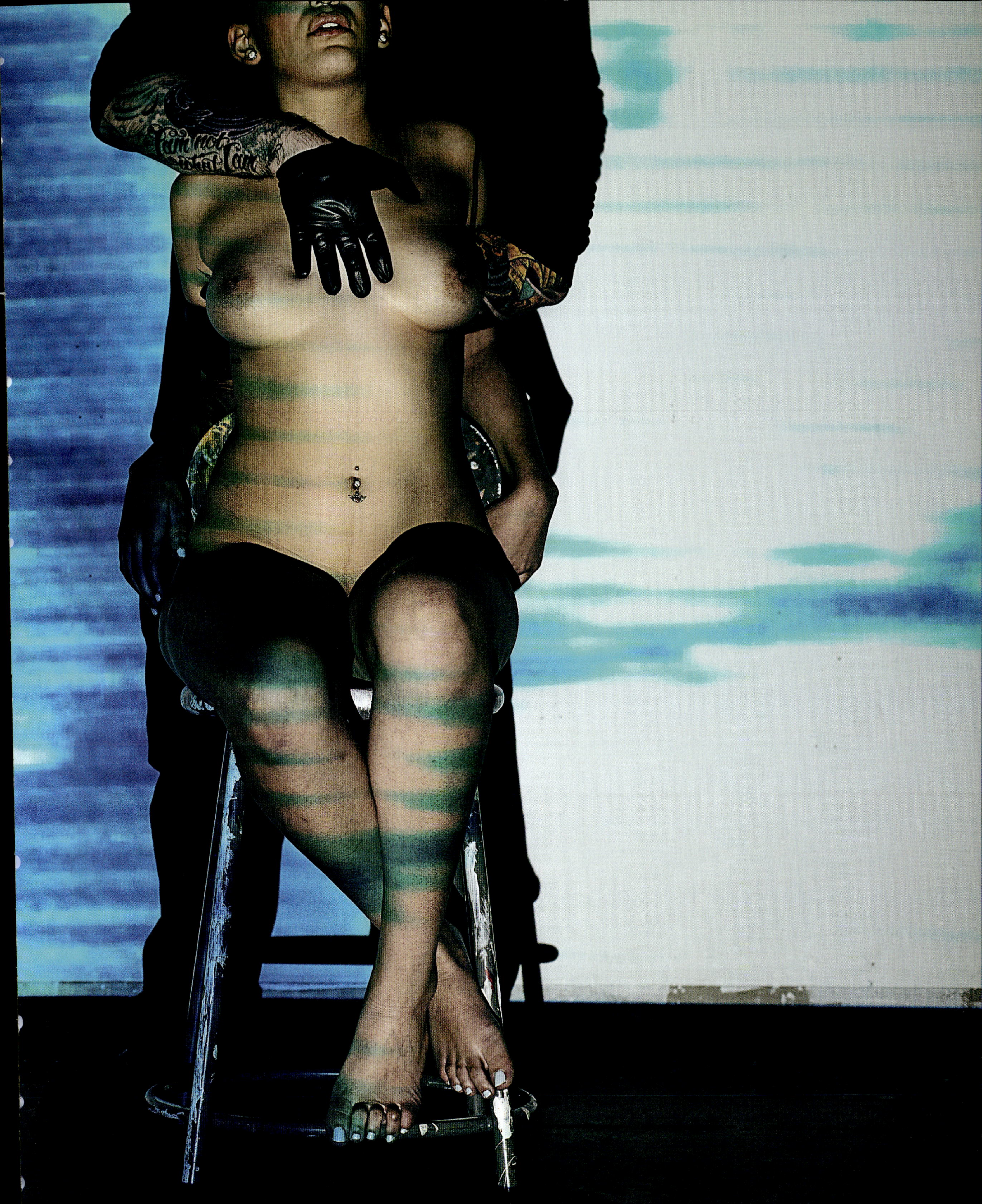

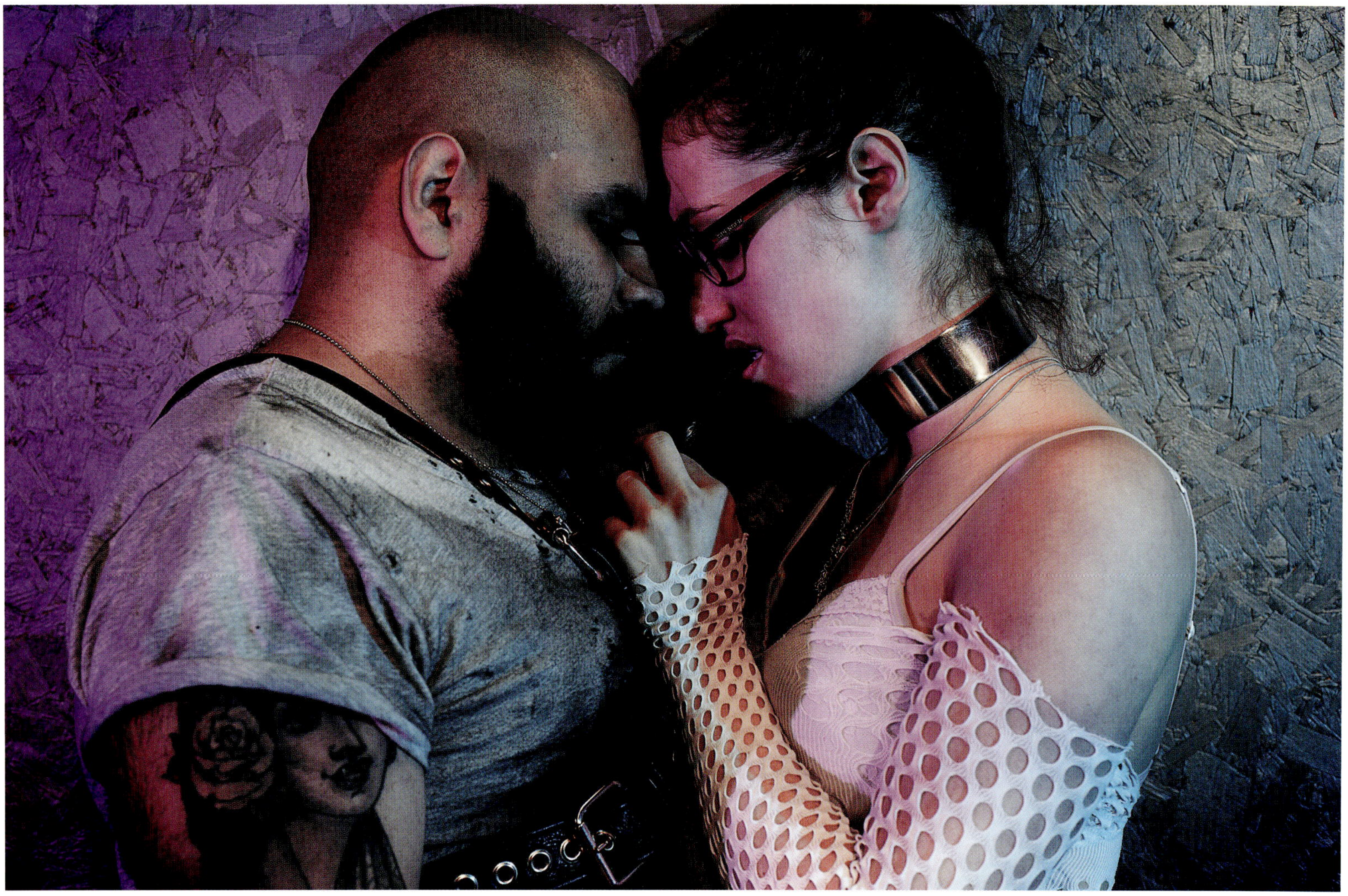

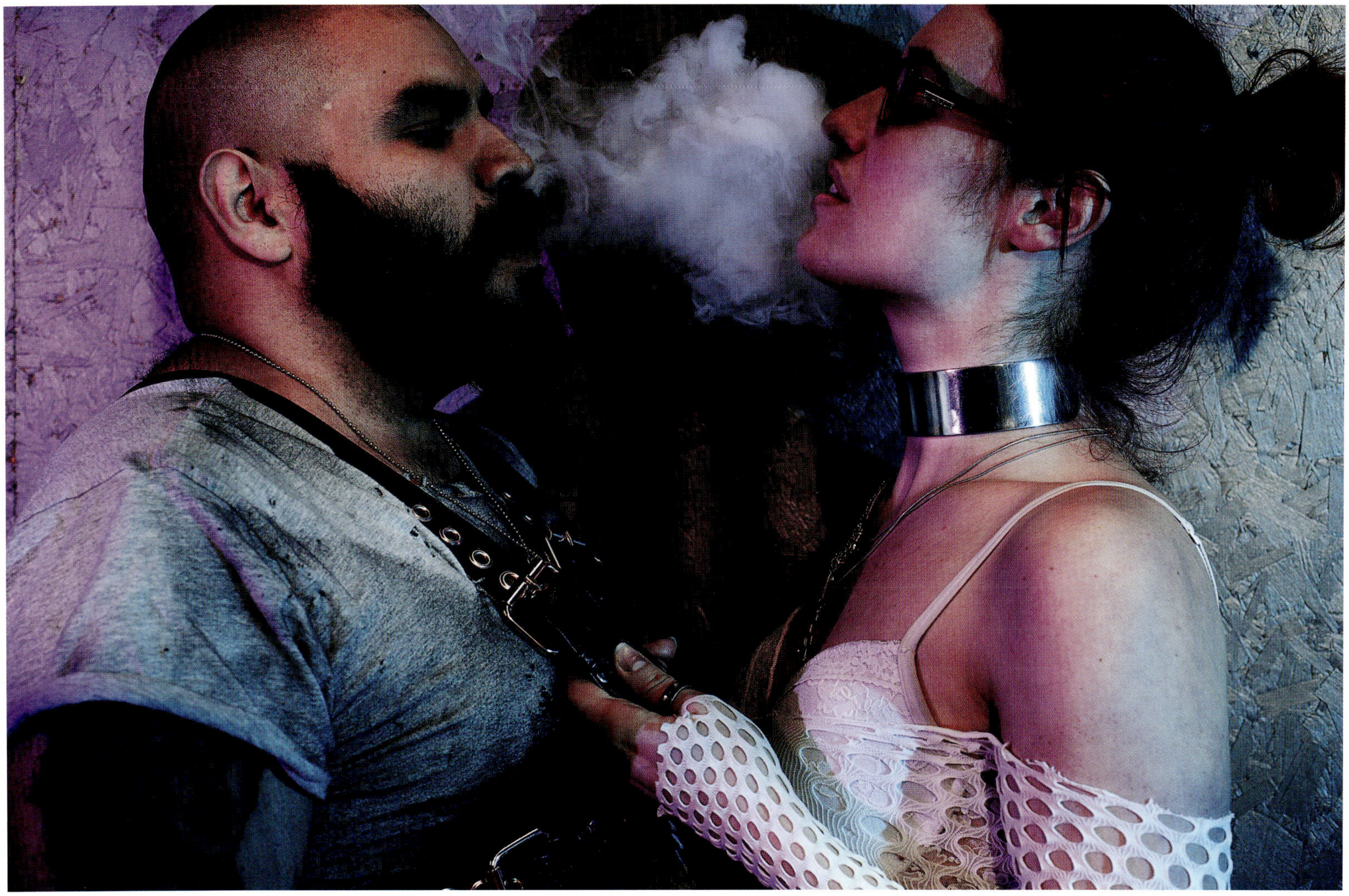

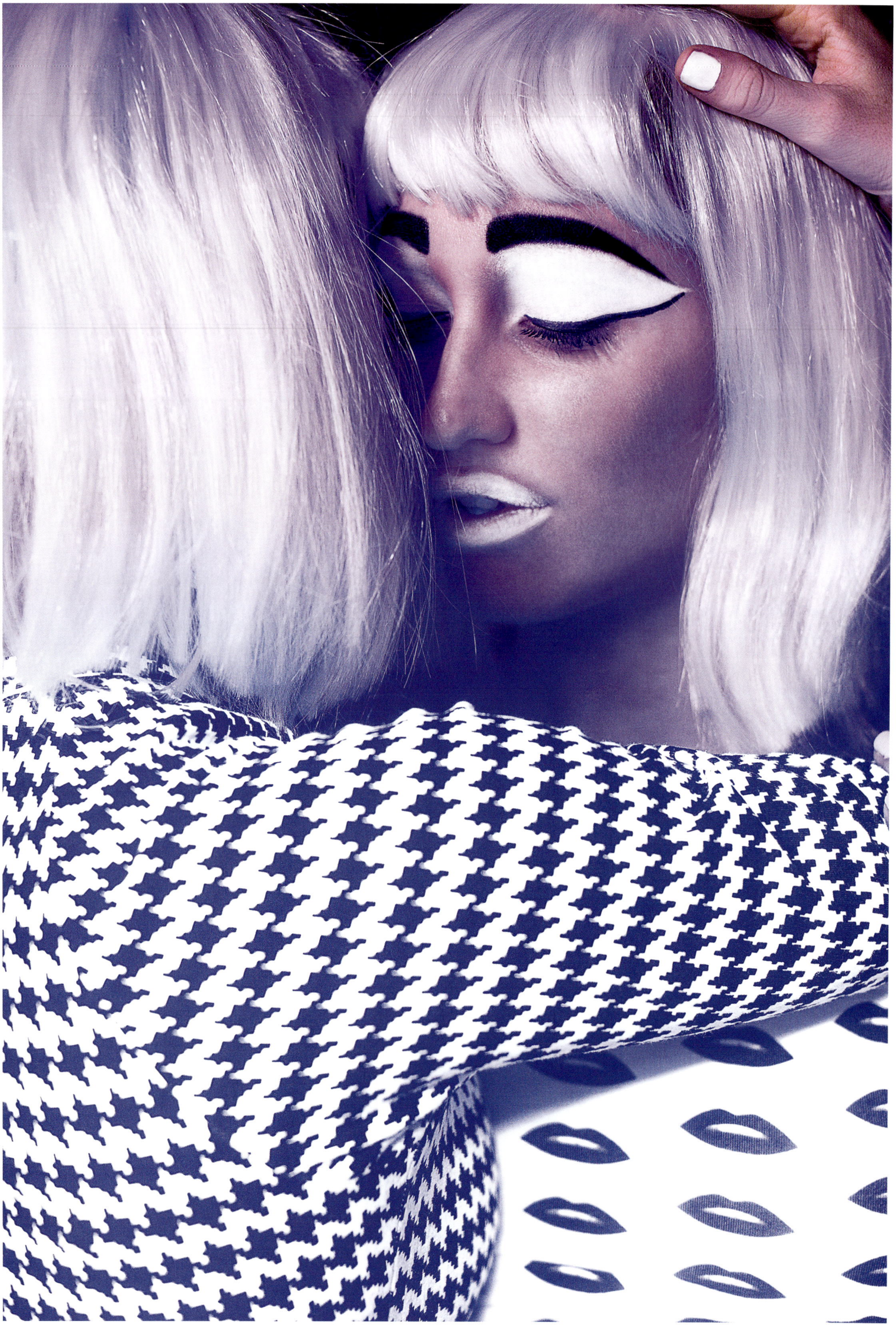

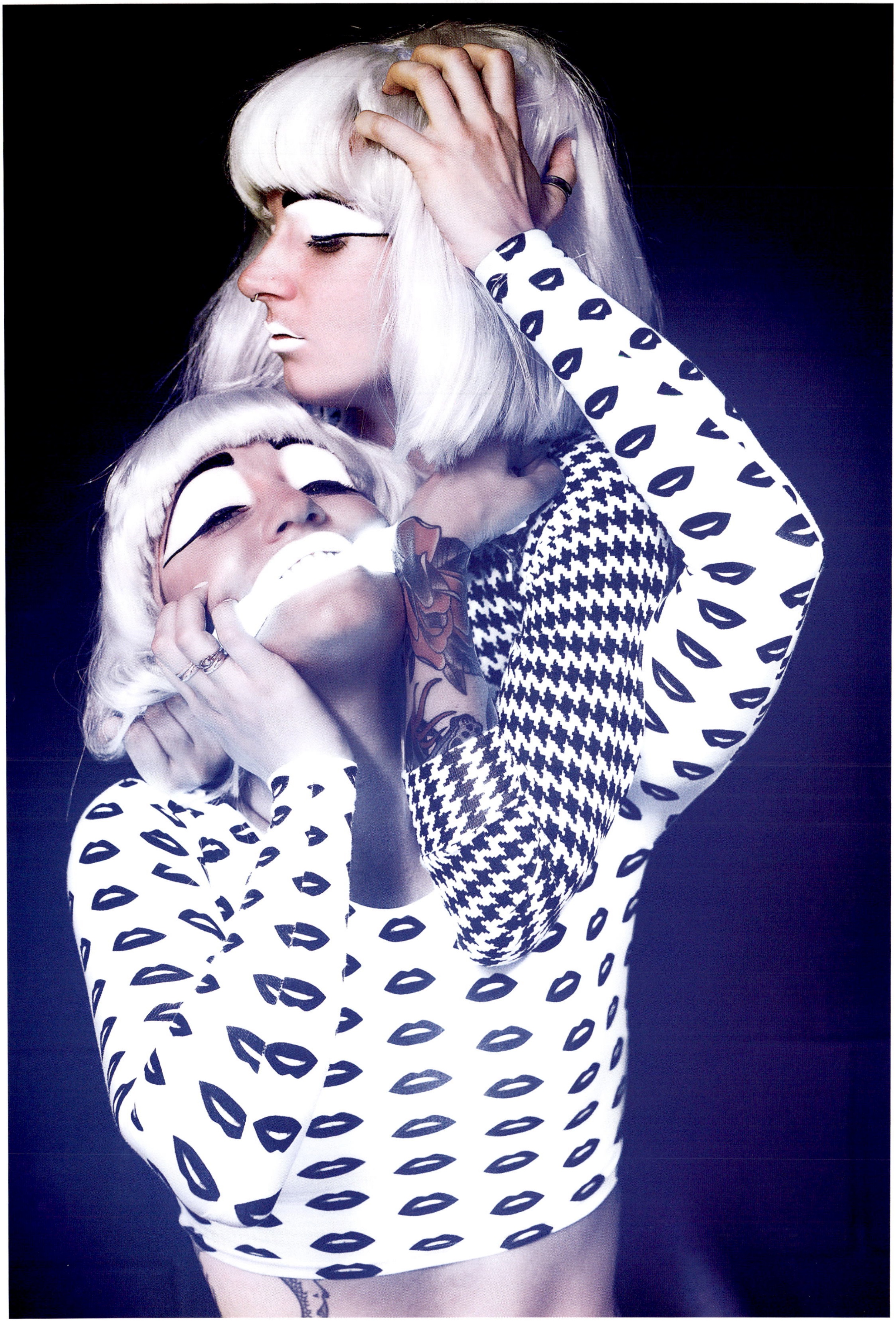

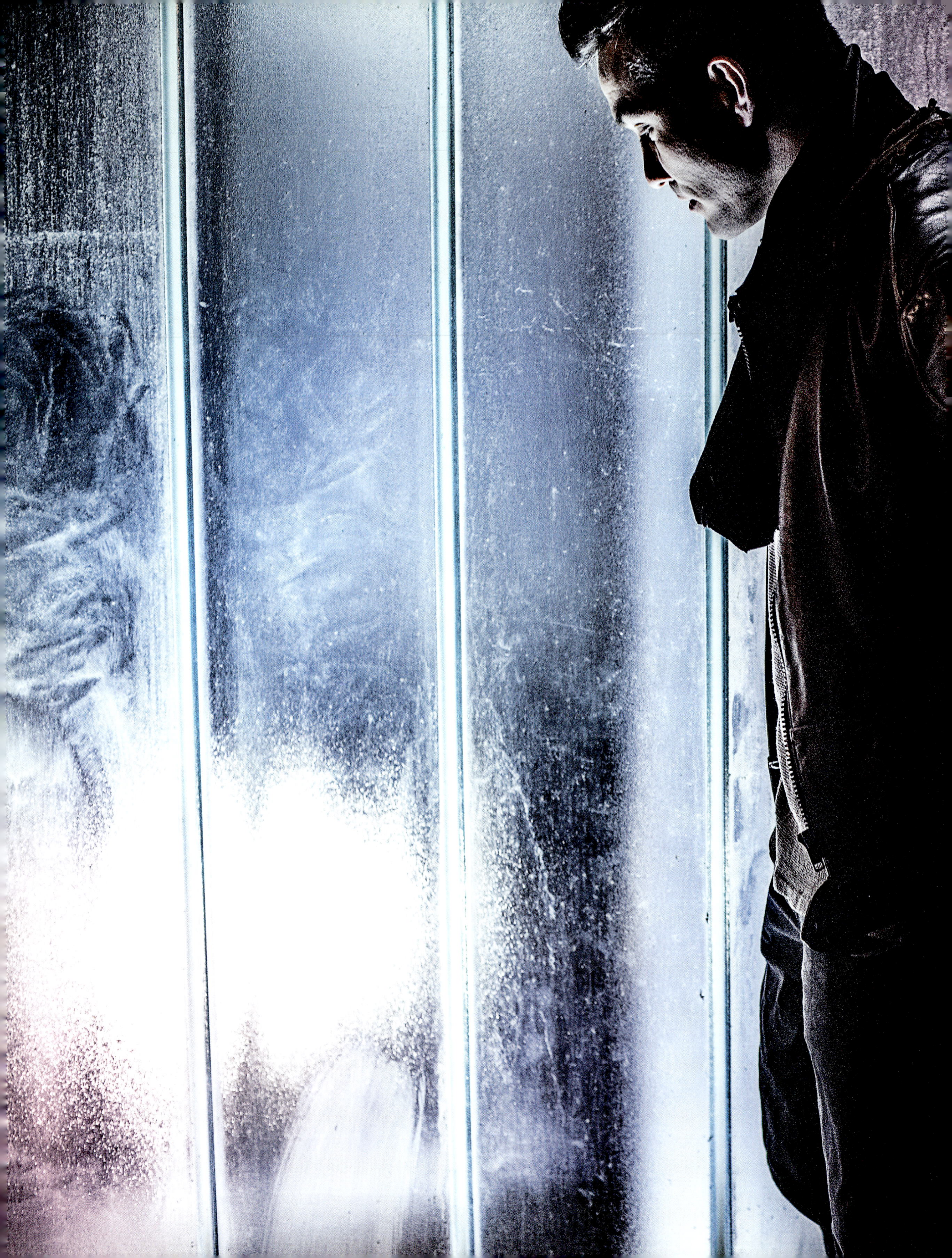

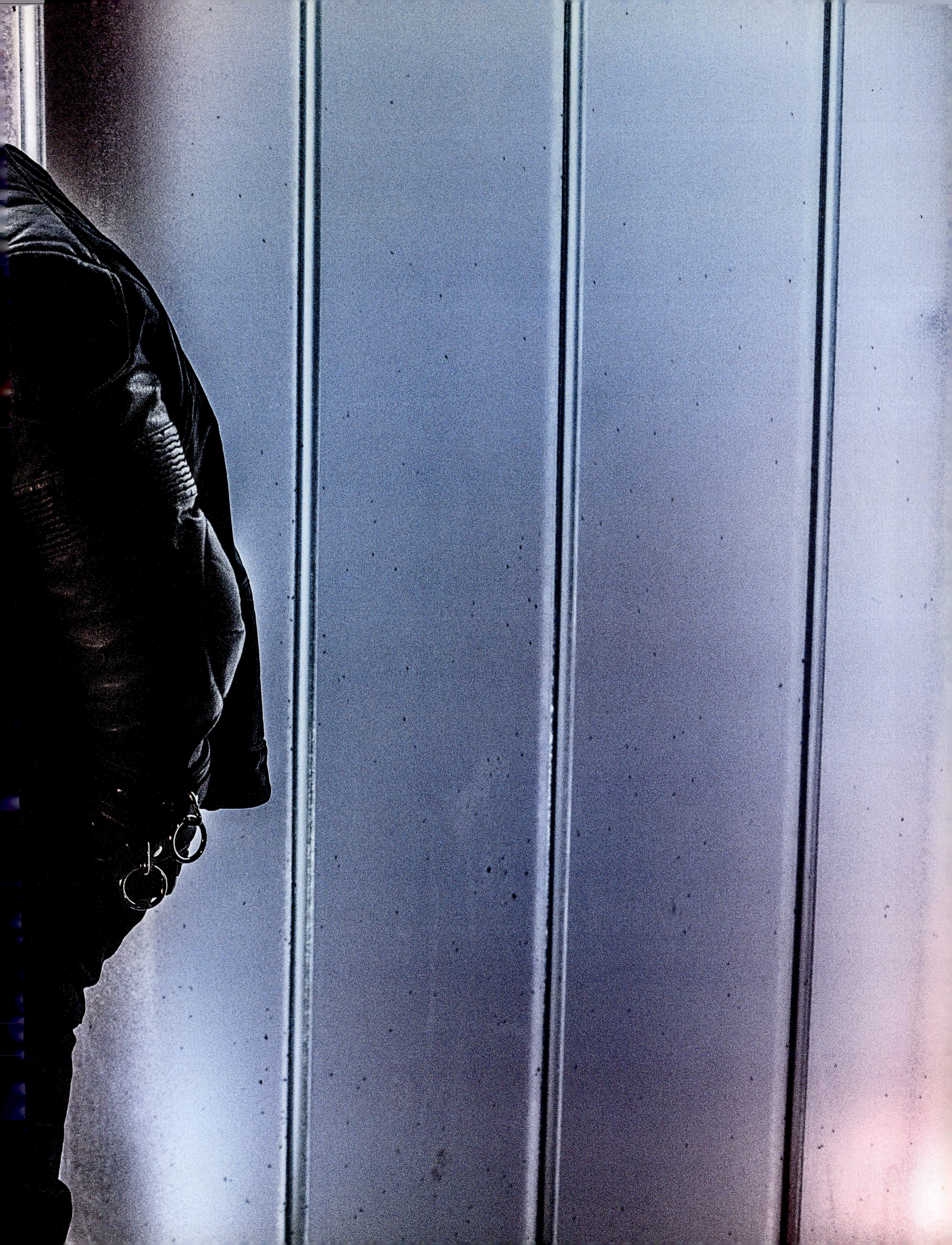

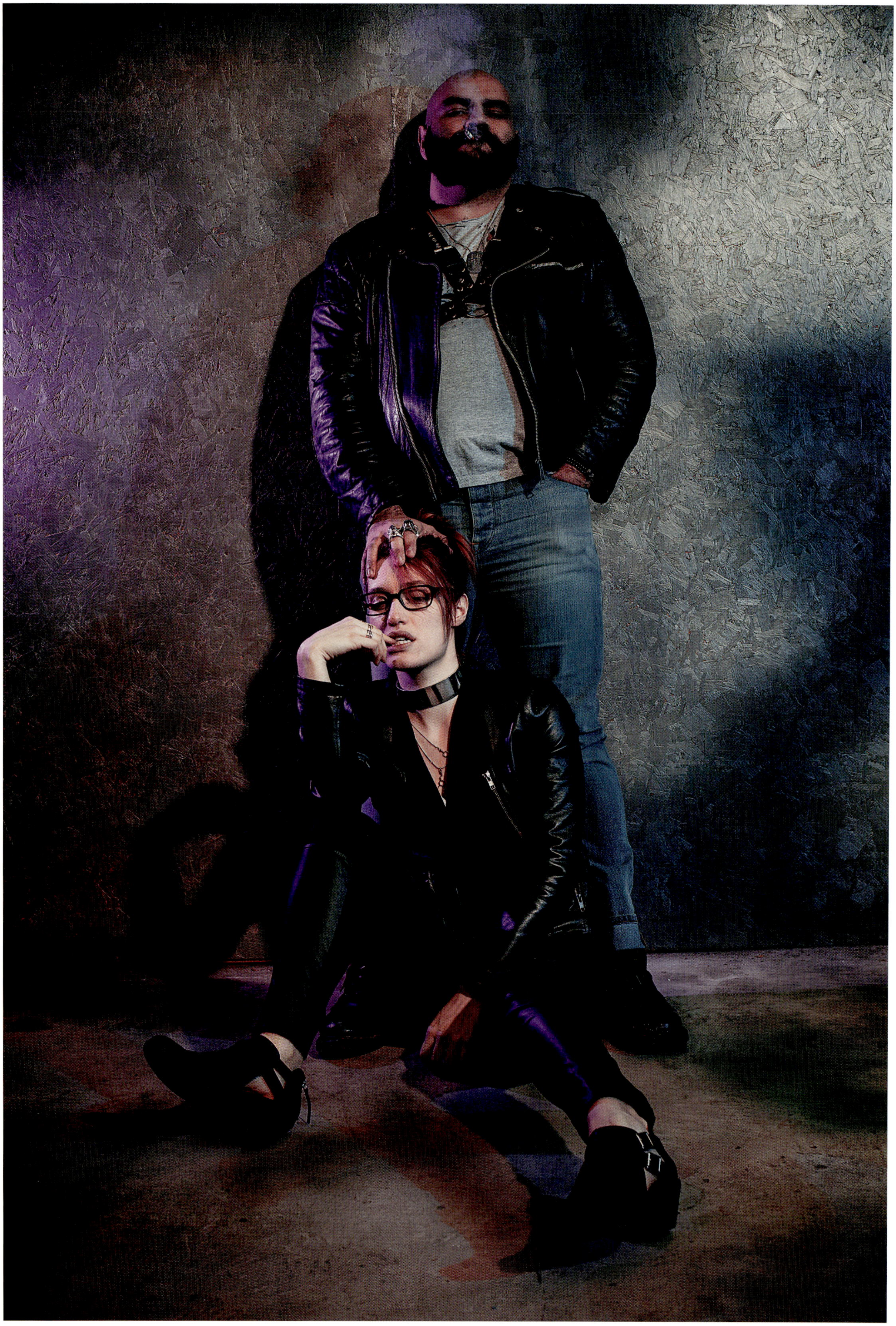

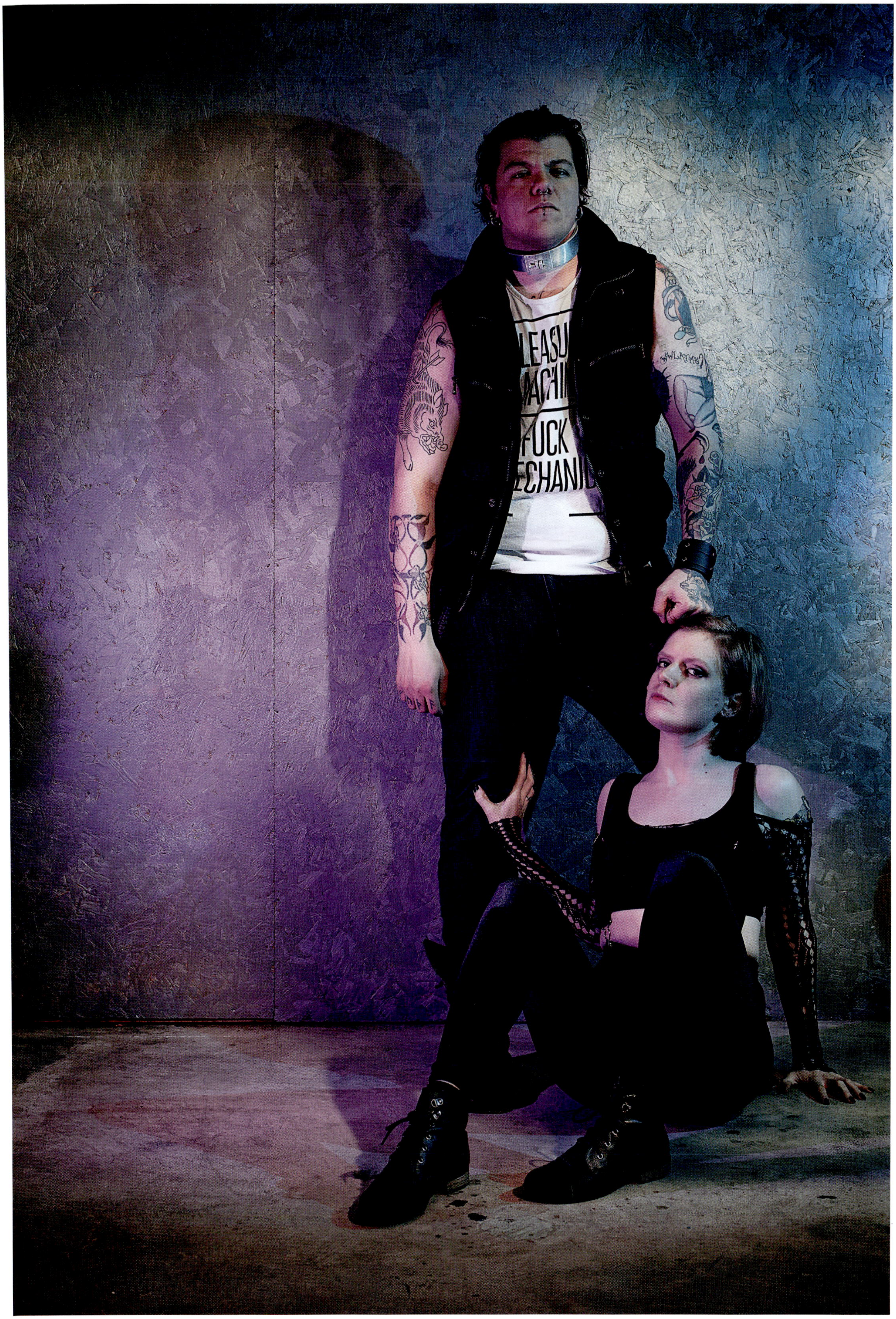
LEASU
MACHI
FUCK
ECHANIC

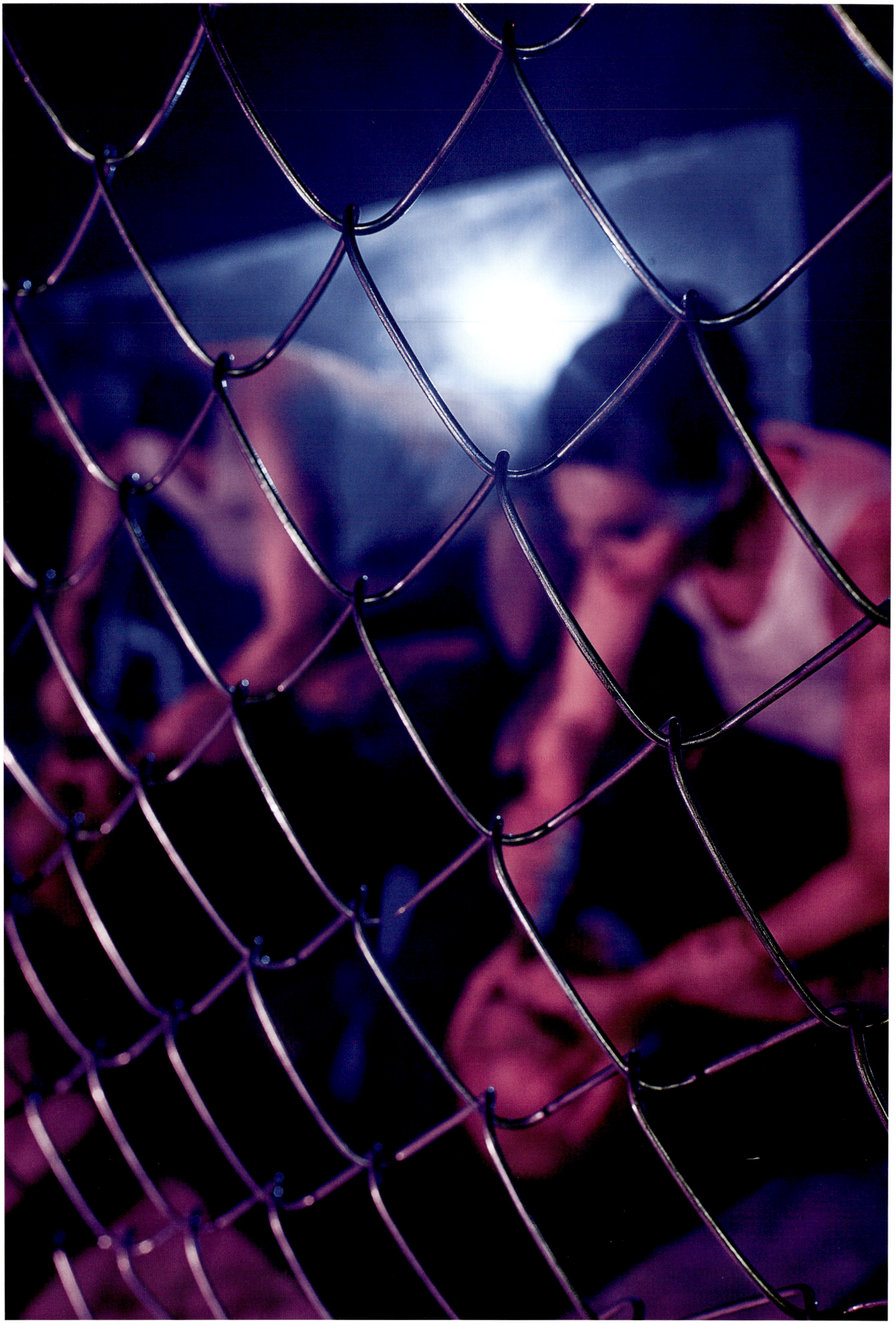

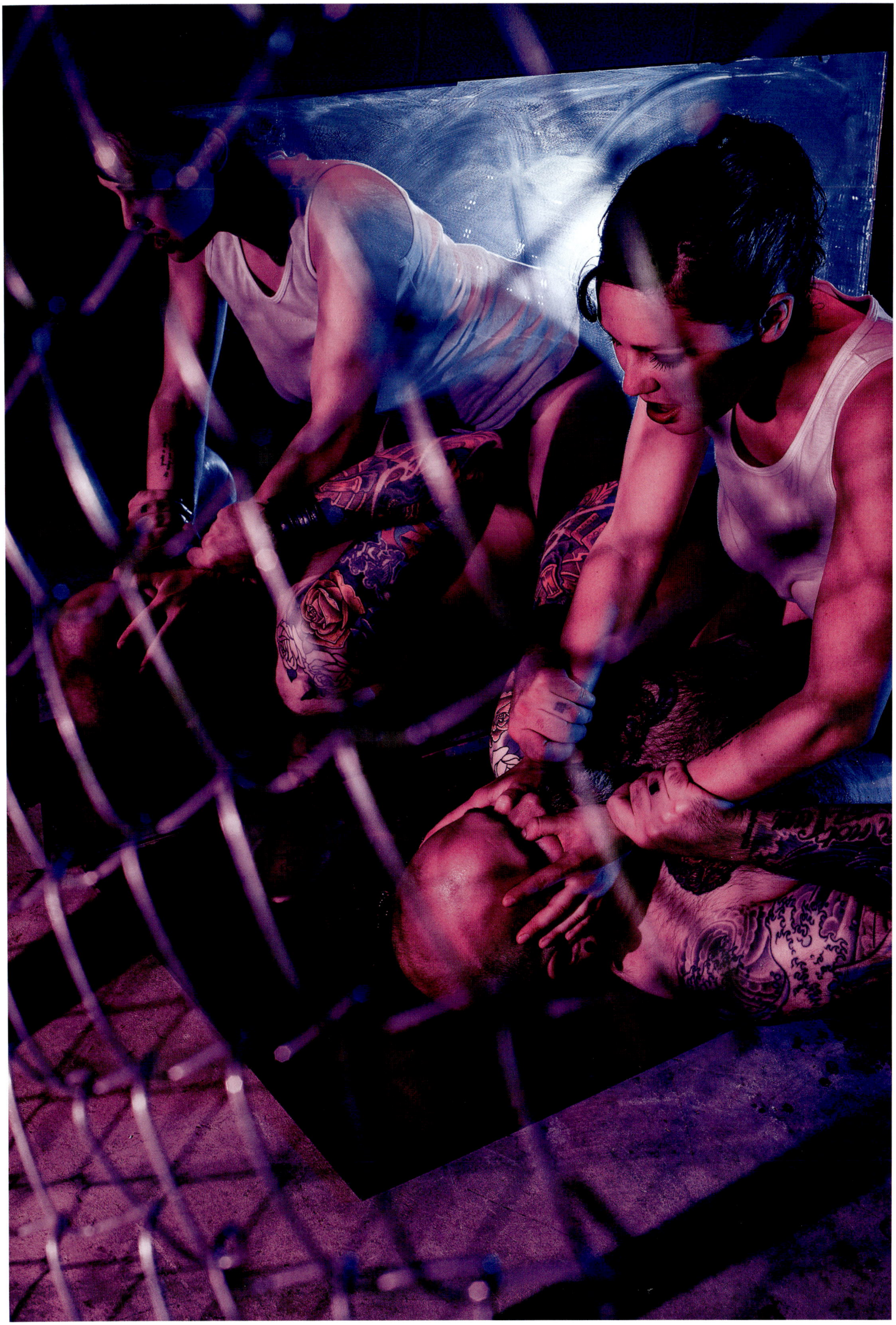

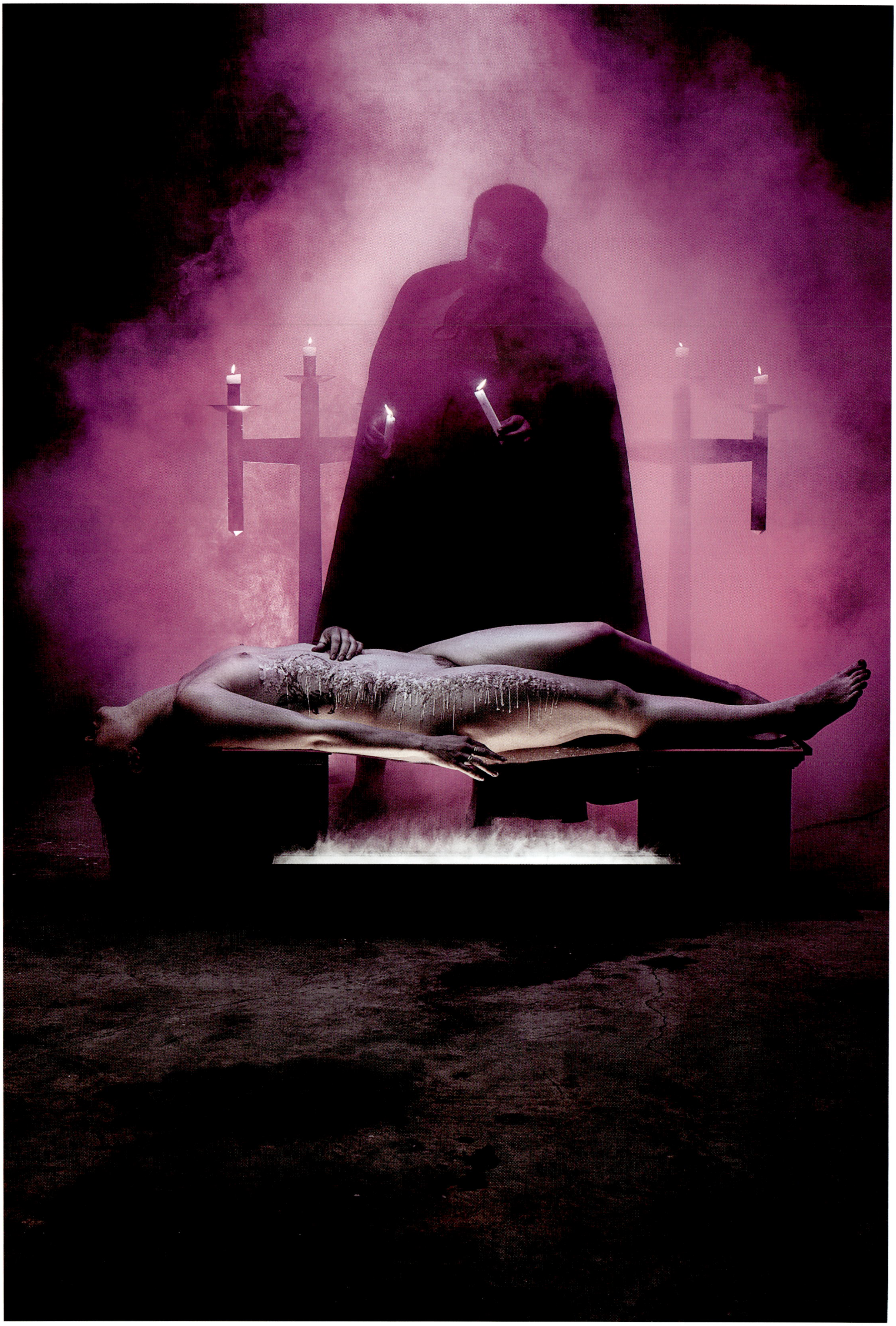

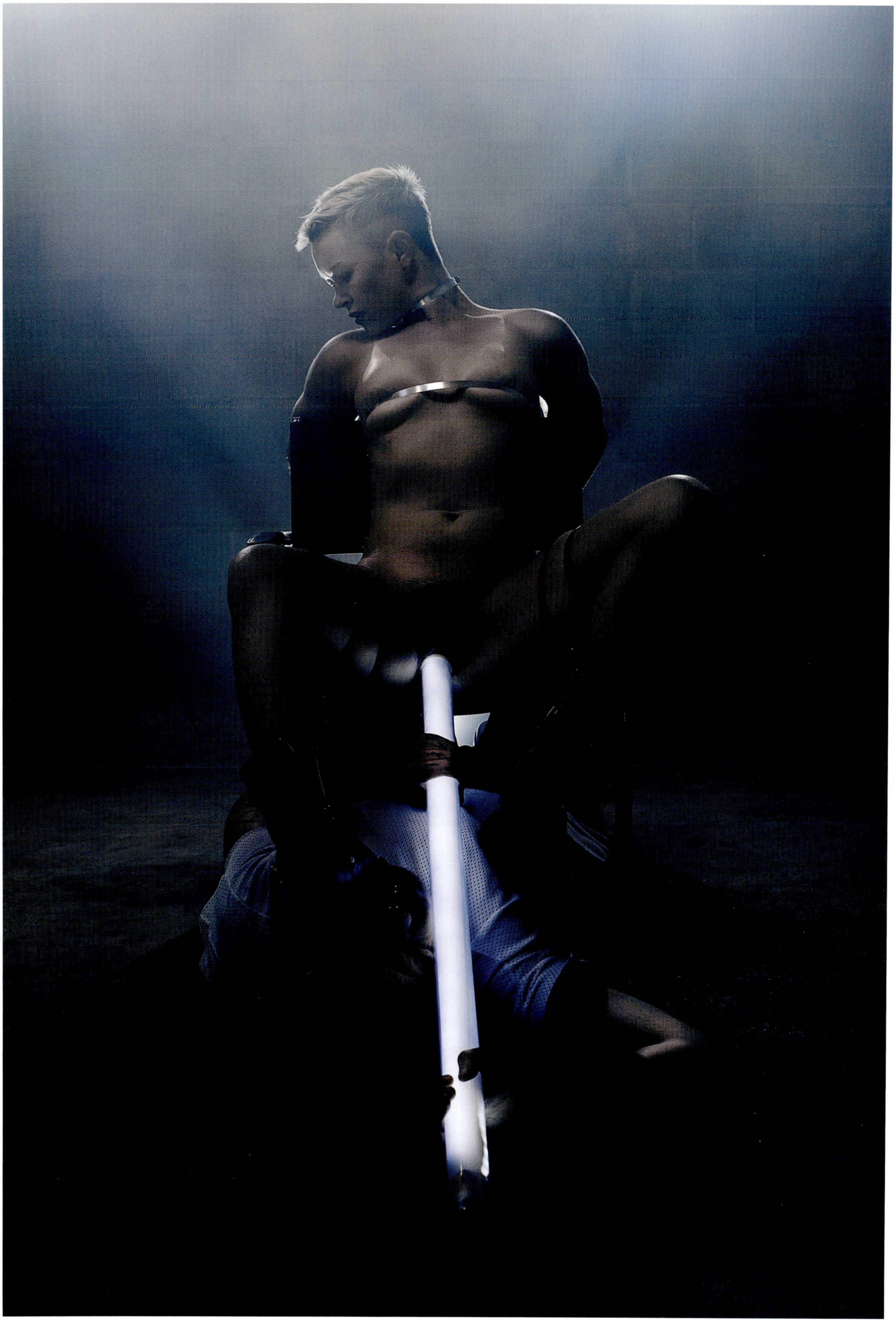

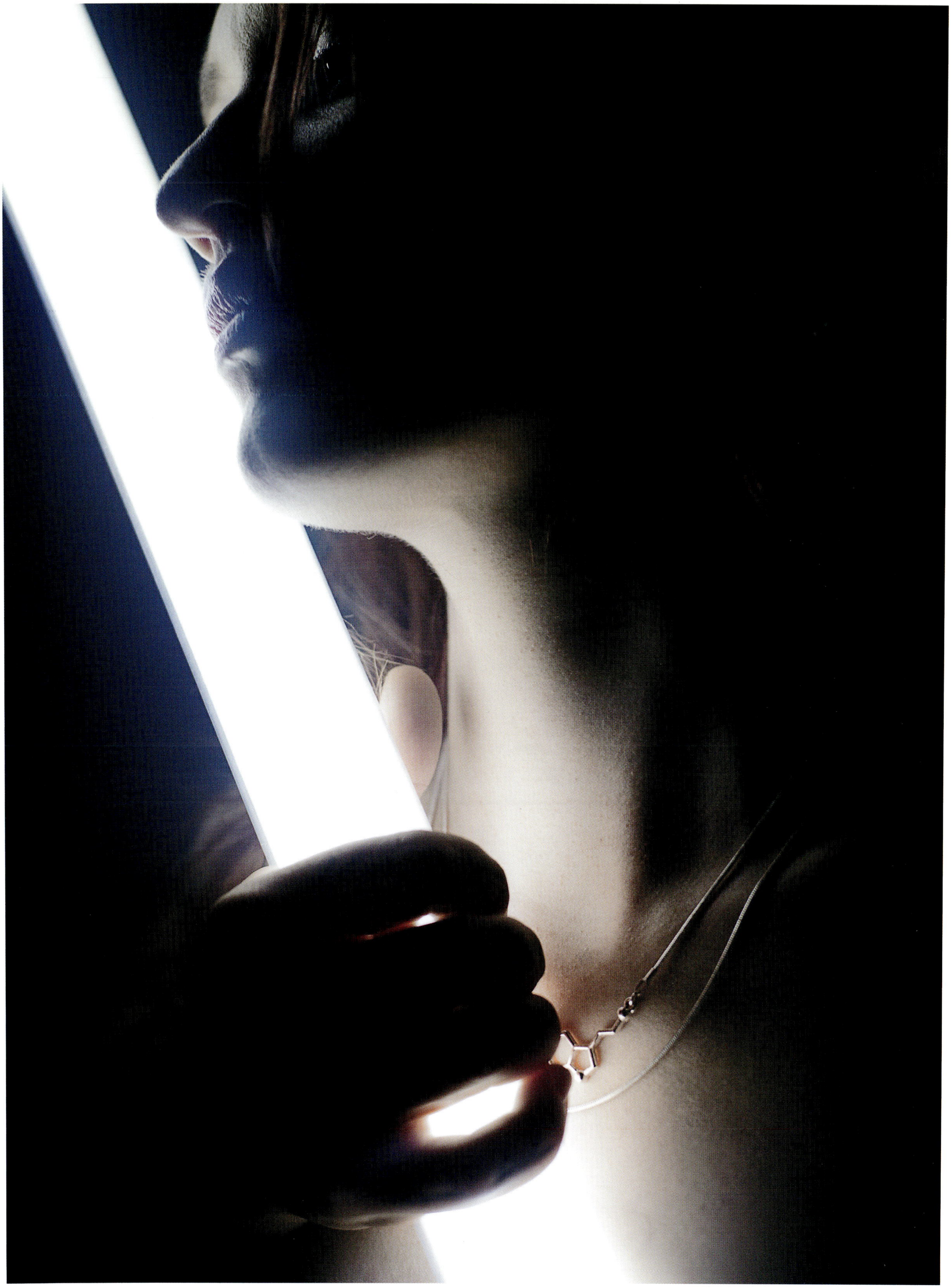

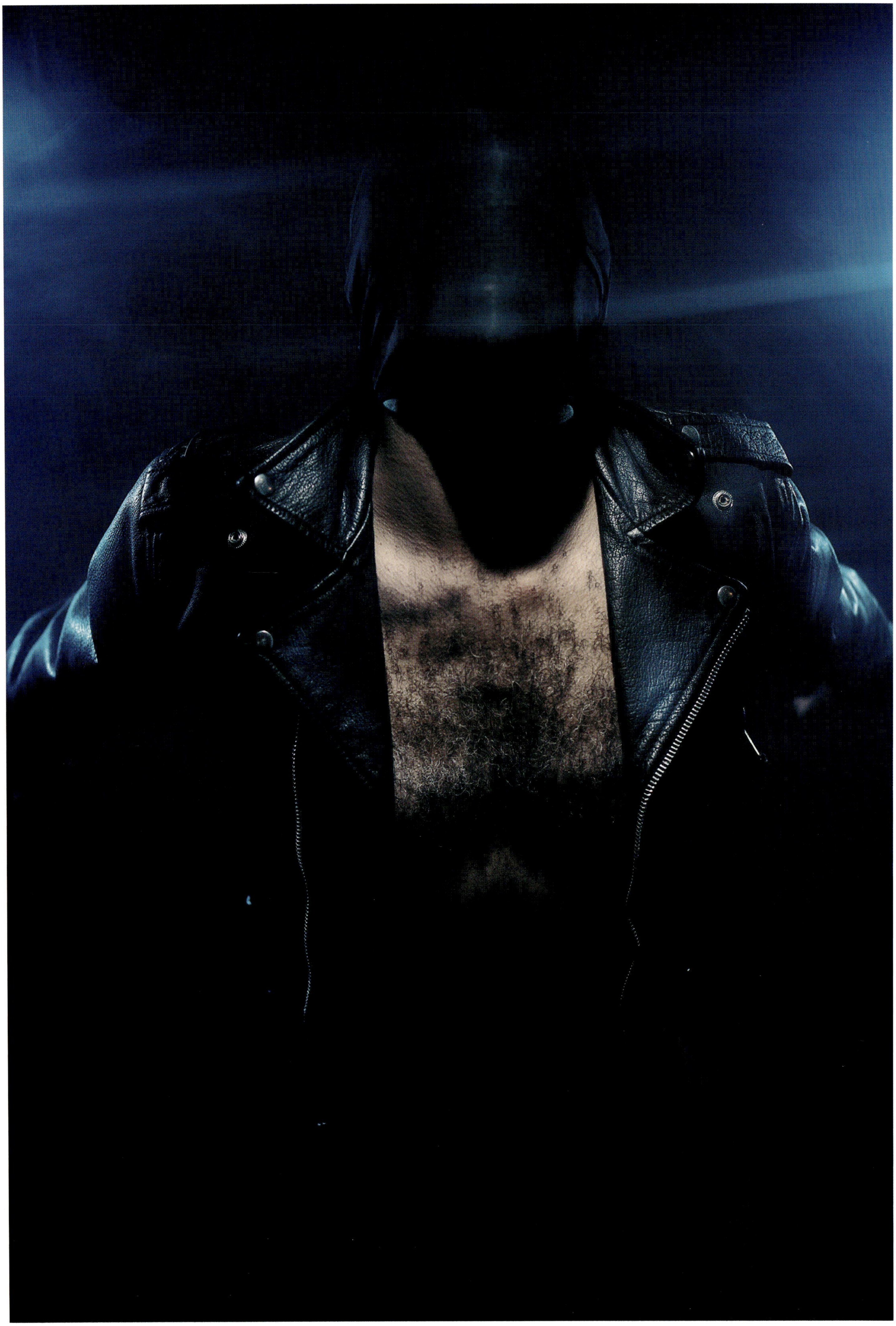

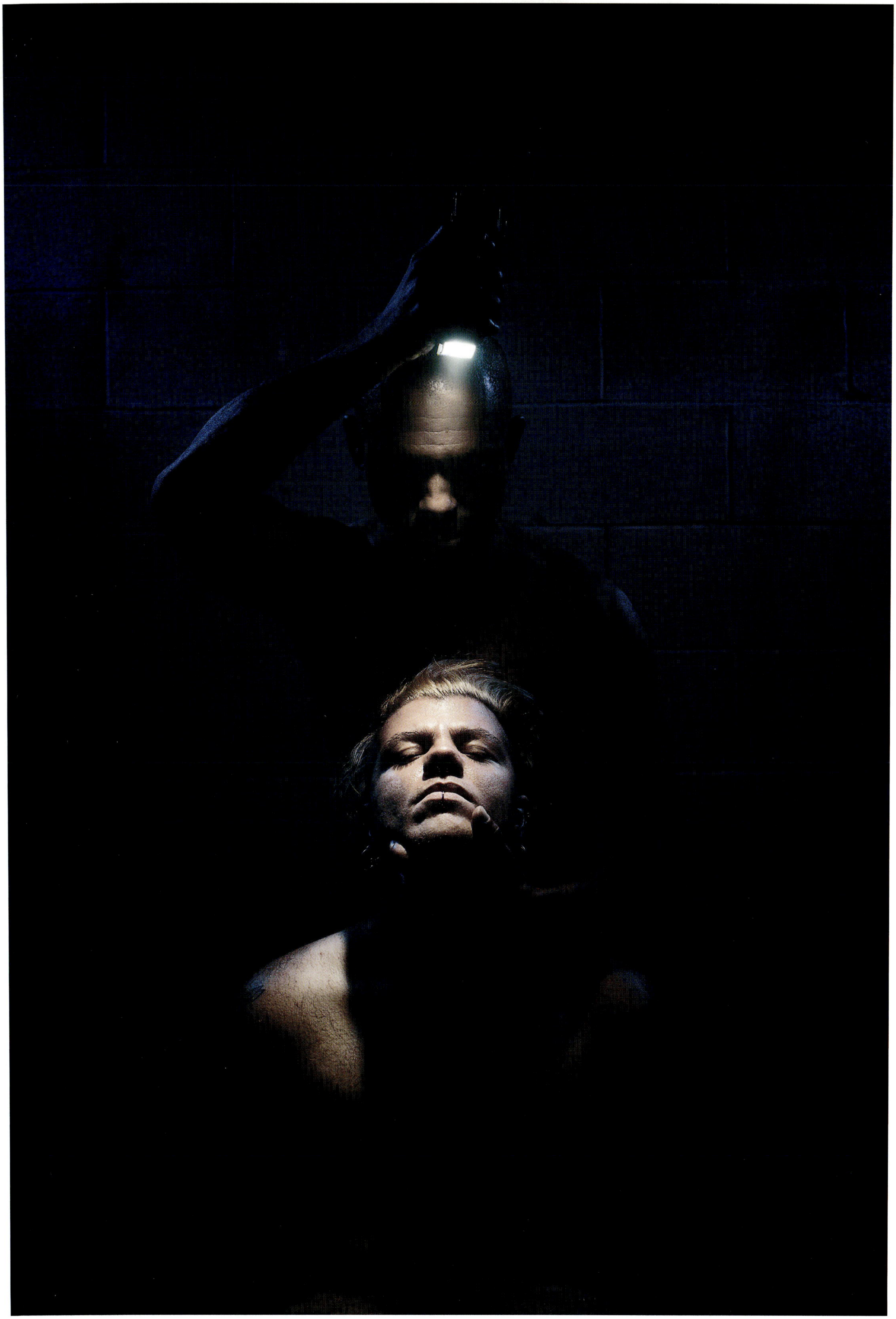

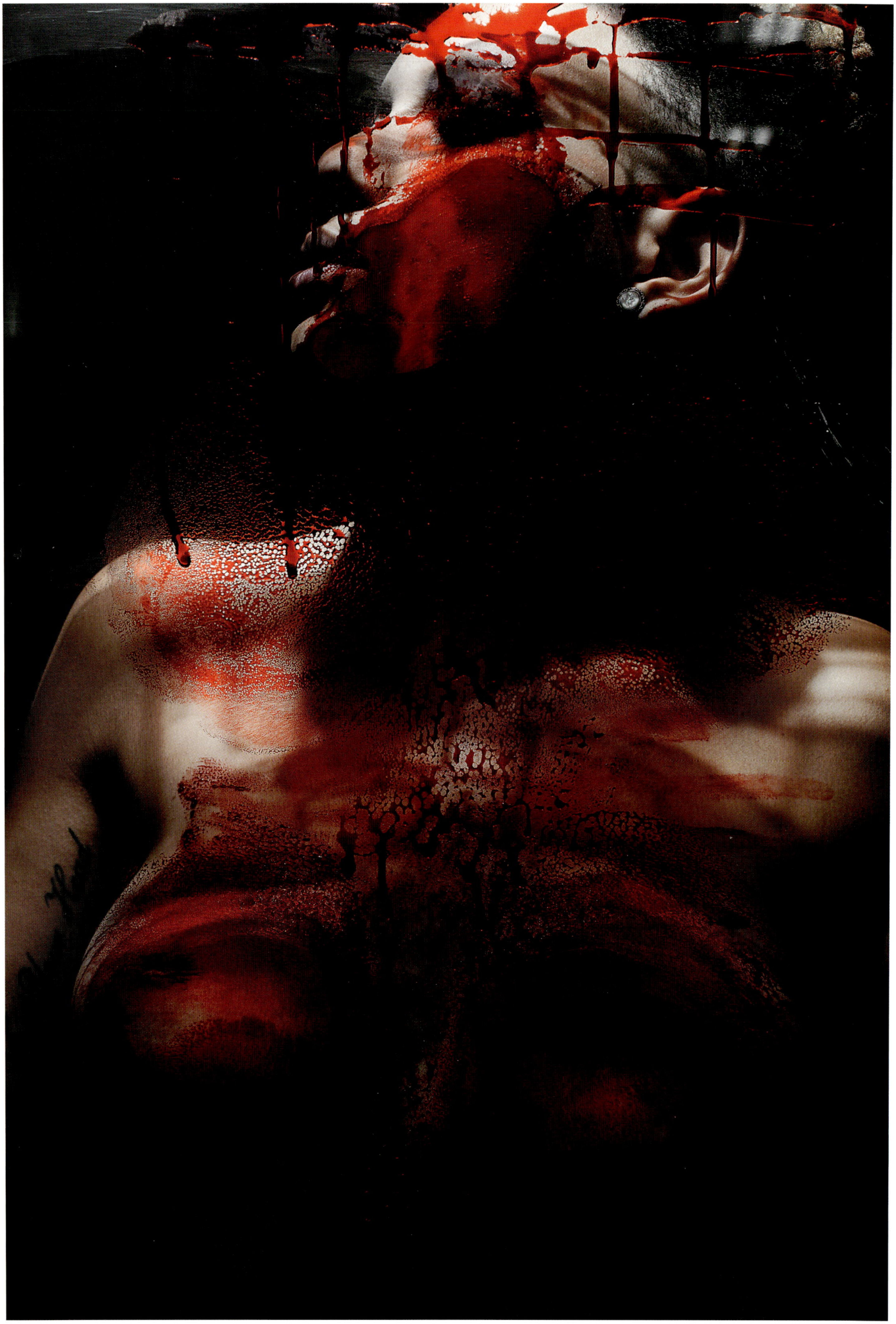

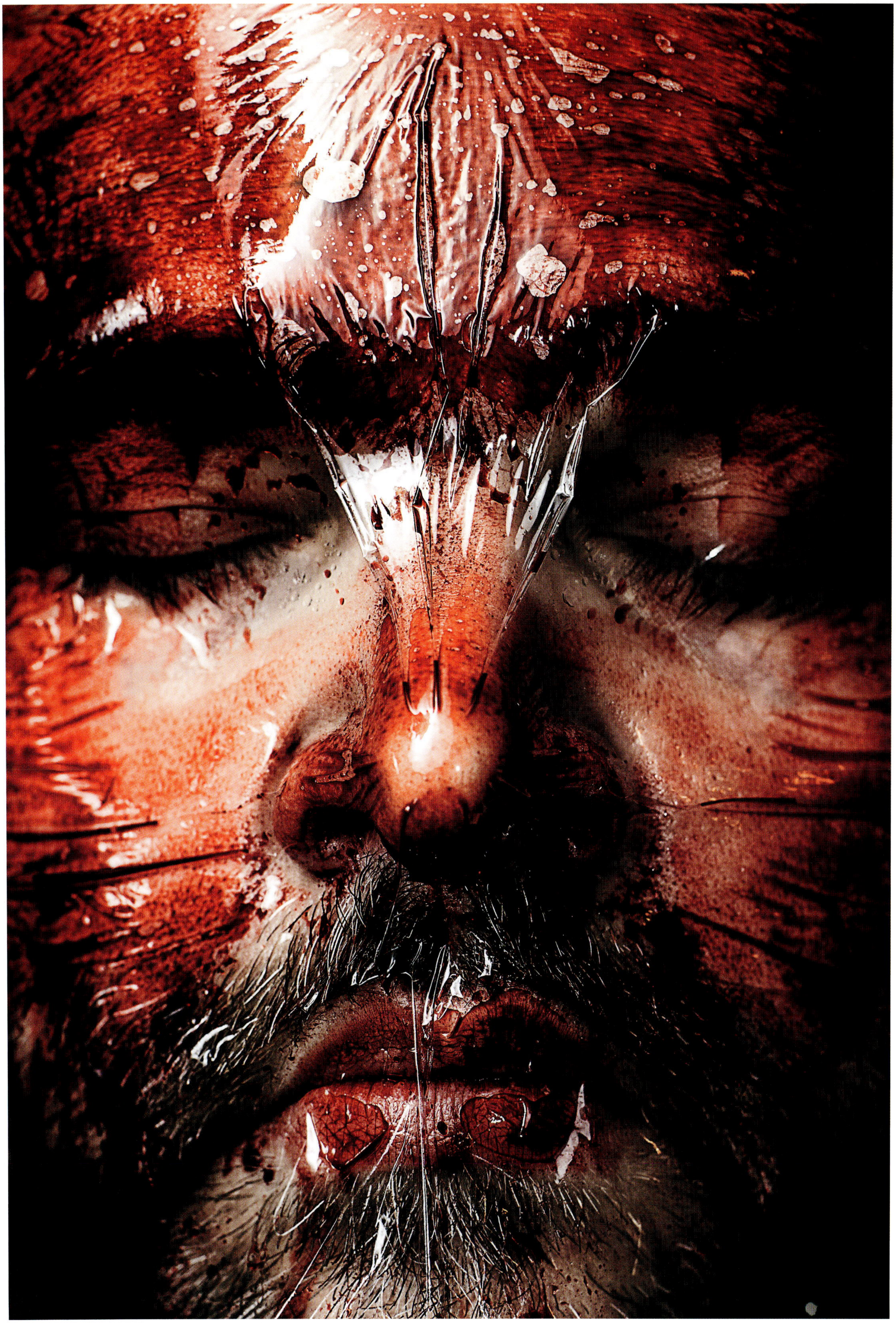

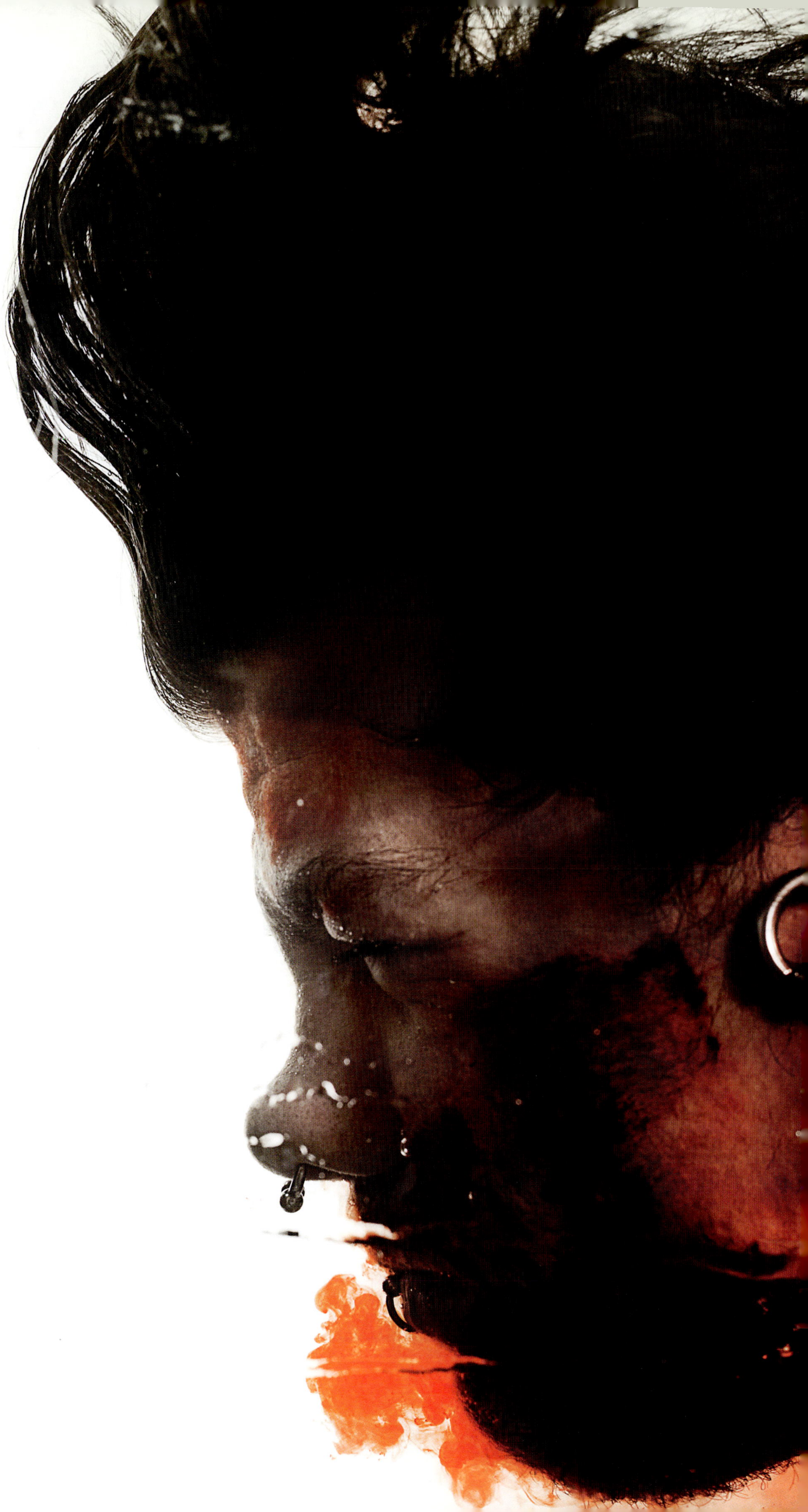

To the 7 unreasonably good looking folks who said yes when I gave them a vague if not non existent description of what I would be asking them to do, you all fucking rock. You showed up last minute, when I would get some weird idea into my head and trusted that somehow I would make something good come of it even though your bodies were covered in some strange substance and that smell would be haunting us for days (note to all - skim milk powder is fucking gross).

To Tina - who made all of this a reality - you are a wonder and I always feel like the luckiest to have you on my side.

To Zach - who made the cover so much more badass than I could've done - I'll flaunt this around when you're too famous to talk to me.

To Dan - thanks for being my gay BearDad and hugging me when that fucker went up my neck.

To Jiz - I can't think of a better person to be stamped into the ground by.

To David - thanks for helping set the tone for the entire project, and for carying the dirt all the way through to the end.

To the mother - who I know would drive back across the country to murder me if I left her out - thanks for not sending 8 year old me back to bed when I *thought* I was being sneaky and watching Bound through the slats of the stairwell. And, you know, for the lifetime full of love and support. Most of us don't get badass, bat shit crazy, radical queer femme dyke moms, and I wouldn't trade it for all the stability in the world.

To Dad - thanks for all the fish (and for making sure I wasn't the only nerd in the family).

To the Minion - who made things that are real pretty and in turn made my photos prettier - thanks for being a voice activated shutter robot, and for countless late nights and for reminding me that I am good at things and for encouraging my already planet sized ego and and and.

To Leitch - who really fucking gets it - thank you for being so good at words when I couldn't be, and for picking out all the easter eggs on the cover without taking a breath - in all honesty, you make me feel like _not_ a terrible human being.

To the Wife - even though we couldn't do this thing together, I'm never not aware that the person I am today could not have been without having been with you.

To Ivory - who delivered a coffee care package to my mailbox while I was sleeping the day before this was due, knowing I didn't have any to wake up to - you are a fucking beast of a human, and I would not have survived the last 6 months (let alone finished this) without you strong arming and whiskeying me through it.

To the Half-Max, who has spent the last decade+ being down for any random thing in the name of art and fun - thank you for taking me to see Robyn so that we could dance around discussing your reactions to masked aggressive men. You complete me.

Half-Max, JP, German, Nomi, Wil, Skye, Everly - We made a badass thing in the world and even if it went nowhere, it is the proudest I have even been of a thing.

I'M DONE BEING SQUISHY NOW - TAKE OFF YOUR CLOTHES

XXX -M

CREDITS

PHOTOGRAPHY - N MAXWELL LANDER

PHOTO ASSISTANT - IZZIE COLPITTS CAMPBELL

COVER DESIGN - ZACHARY J BAUTISTA

FOREWORD - ALEX LEITCH

CAST

JP ROBICHAUD @CONTROLENTHUSED

GERMAN SHIBLE @GERMANSHIBLE

SKYE CHEVOLLEAU @SKYECHEVOLLEAU

EVERLY FOX @EVERLYFOX

HALF-MAX

WILLIAM ALEXANDER LEE

NAOMI LANG

N MAXWELL LANDER

CAMEOS

DAN HENDERSON @VALHALLA_IMAGES

JIZ LEE @JIZLEE

IVORY @PUREIVORYDOTCA

DAVID FINDLAY

THE END

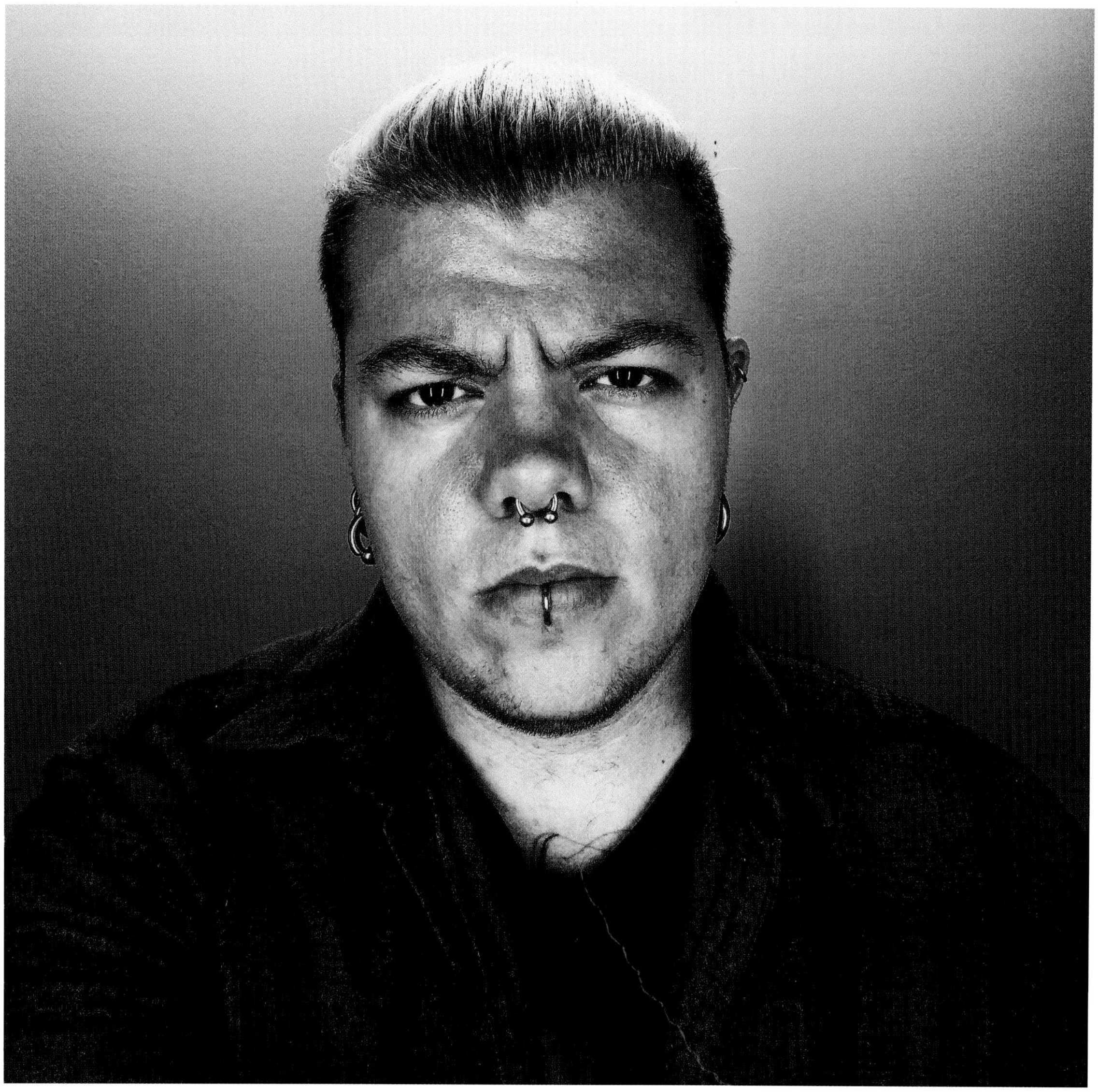

N Maxwell Lander is a photographer, designer, and other-type-of-artist.

He is an advocate of the Oxford Dictionary including "pig" as a classification of gender and in an ideal world would only be allowed to respond with "aspiring bearmanthing" when asked if he was a boy or a girl.

He has three life mottos - "Don't get a dog with eyebrows," "Yes, we do need more fake blood in the budget," and "I'm pretty sure you can come again."

He remembers being told his first word was "chocolate" and that his first steps were taken in order to get closer to certain parts of Dolly Parton on a TV screen.

He aspires to someday grow a beard and hopes to use his newly acquired deep voice to become the fat guy in the dance music video that inserts the singer's name into the beginning of the track.

MAXWELLANDER.CA
@MAXWELLANDER